Images and Artefacts of the Ancient World

THE BRITISH ACADEMY · THE ROYAL SOCIETY

Images and Artefacts of the Ancient World

Edited by
Alan K. Bowman & Michael Brady

Published for THE BRITISH ACADEMY
by OXFORD UNIVERSITY PRESS

Oxford University Press, Great Clarendon Street, Oxford OX2 6DP

Oxford New York
Auckland Bangkok Buenos Aires Cape Town Chennai
Dar es Salaam Delhi Hong Kong Istanbul Karachi Kolkata
Kuala Lumpur Madrid Melbourne Mexico City Mumbai Nairobi
São Paulo Shanghai Singapore Taipei Tokyo Toronto

Oxford is a registered trademark of Oxford University Press
in the UK and certain other countries

Published in the United States
by Oxford University Press Inc., New York

British Library Cataloguing in Publication Data
Data available

ISBN 0–19–726296–1

Typeset by Latimer Trend & Company Ltd
Printed in Great Britain
on acid-free paper by
Henry Ling Ltd, at the Dorset Press, Dorchester, Dorset

This is a British Academy Occasional Paper, No. 4

Contents

List of Figures and Table   vi

Notes on Contributors   x

Preface   xiii

Introduction   1
    MICHAEL BRADY & ALAN K. BOWMAN

1. Wooden Stilus Tablets from Roman Britain   7
    ALAN K. BOWMAN & ROGER S. O. TOMLIN

2. Shadow Stereo, Image Filtering, and Constraint Propagation   15
    MICHAEL BRADY, XIAO-BO PAN, VEIT SCHENK, MELISSA TERRAS,
    PAUL ROBERTSON & NICHOLAS MOLTON

3. Digitising Cuneiform Tablets   31
    CARLO VANDECASTEELE, LUC VAN GOOL, KAREL VAN LERBERGHE, JOHAN VAN ROMPAY
    & PATRICK WAMBACQ

4. Interpretation of Ancient Runic Inscriptions by Laser Scanning   35
    JAN O. H. SWANTESSON & HELMER GUSTAVSON

5. Virtual Reality, Relative Accuracy: Modelling Architecture and Sculpture with VRML   45
    MICHAEL GREENHALGH

6. Automatic Creation of Virtual Artefacts from Video Sequences   59
    ANDREW W. FITZGIBBON, GEOFF CROSS & ANDREW ZISSERMAN

7. At the Foot of Pompey's Statue: Reconceiving Rome's *Theatrum Lapideum*   69
    HUGH DENARD

8. Modelling Sagalassos: Creation of a 3D Archaeological Virtual Site   77
    LUC VAN GOOL, MARC POLLEFEYS, MARC PROESMANS & ALEXEY ZALESNY

9. Three-Dimensional Laser Imaging in an Archaeological Context   89
    ANDREW M. WALLACE

10. Movements of the Mental Eye in Pictorial Space   99
    JAN J. KOENDERINK

11. The Potential for Image Analysis in Numismatics   109
    CHRISTOPHER J. HOWGEGO

12. Italian *Terra Sigillata* with Appliqué Decoration: Digitising, Visualising,
and Web-Publishing   115
    ELENI SCHINDLER KAUDELKA & ULRIKE FASTNER

13. Shape from Profiles   125
    ROBERTO CIPOLLA & KWAN-YEE K. WONG

14. The Skull as the Armature of the Face: Reconstructing Ancient Faces   131
    R. A. H. NEAVE & A. J. N. W. PRAG

15. Reconstruction of a 3D Mummy Portrait from Roman Egypt   145
    ALF LINNEY, JOÃO CAMPOS & GHASSAN ALUSI

List of Figures and Table

Figures

1.1. *Tab. Vindol.* I 25 = II 247   7
1.2. Vindolanda stilus tablet 797   10
1.3. Vindolanda stilus tablet 836   10
1.4. Vindolanda stilus tablet 974   11
1.5. Tablet 974, drawn by R. S. O. Tomlin   12
1.6. Tablet 974: table of letter forms, drawn by R. S. O. Tomlin   12
2.1. Vindolanda stilus tablet 974, with magnified portion   15
2.2. Two images of a fragment of a stilus tablet   17
2.3. Geometry of shadow formation   18
2.4. Thresholded correlation of phase congruency (PC) feature values   19
2.5. Wood-grain removal   19
2.6. A fragment of a stilus tablet with shadows and highlights   20
2.7. Log-Gabor functions of different bandwidths and centre frequencies   21
2.8. Phase congruency and phase of a sample signal   22
2.9. Phase angles of different feature type   22
2.10. Detecting features using the phase angle   23
2.11. Idealised test image, with phase congruency and orientation vectors   24
2.12. Phase congruency at different scales   25
2.13. Image of stilus tablet with illumination at 0°   26
2.14. Gaps in the strokes incised on a stilus tablet   26
2.15. The geometry of filling the gap between two stroke portions   26
2.16. Analysis of the different 'reading levels' used by the papyrologists   27
2.17. *Tab. Vindol.* II 255 screenshots   29
3.1. Hand copy of a cuneiform tablet   31
3.2. Two cuneiform tablets, one before and one after chemical treatment   32
3.3. Digital photograph of a cuneiform tablet, with filtered version and image after application   33
4.1. Laser scanner equipment in the laboratory   35
4.2. Sketch map of the locations of runic inscriptions   37
4.3. Contour map of the last rune of the inscription *braido* at Himmelstalund, Norrköping   38
4.4. Part of the vertical row from the inscription at Oklunda   39
4.5. Image of a part of the Röö stone   39
4.6. Digital shadow image of the second half of the inscription on the Skramle stone   40
4.7. Image of the beginning of the two rows on the Kuli stone, Trondheim   41
5.1. VRML (Canoma): a Louvre sarcophagus mapped onto appropriate geometry   48
5.2. Borobudur VRML project: gallery with gauze location overlay   51
5.3. Borobudur VRML project: gallery with gauze overlay for selection   51
5.4. Nineteenth-century panorama: the Colosseum, London   53
5.5. Web imagemap: Heemskerck's panorama of Rome linked to Nolli's map   54
5.6. Web imagemap: the Chigi Chapel, S. Maria del Popolo, Rome   55
5.7. Web imagemap: sculpture of Vishnu   55
5.8. Web imagemap: Christ Church Cathedral, Oxford   55
6.1. Some example sequences   60
6.2. 3D geometry   61
6.3. Projective ambiguity   62
6.4. Pseudocode for truncation of general motion fundamental matrix to single-axis form   62
6.5. Geometry estimation   63
6.6. Track lifetimes   64
6.7. Dinosaur: points and cameras reconstructed from 2D point tracks   64
6.8. Top view of reconstructed cup, with texture-mapped 3D reconstruction   64
6.9. The volumetric constraint offered by a single silhouette   64
6.10. A 2D surface viewed by five 1D cameras   64

6.11. The visual hull from the dinosaur sequence in Figure 6.1   64

6.12. Six silhouette images from a sequence of 120 images of a vase   64

6.13. Images of a skull   66

7.1. Aerial photograph of the area of Rome corresponding to the Theatre of Pompey   70

7.2. Fragments of the Marble Plan   71

7.3. Theatre of Pompey: *porticus post-scaenam* and Temple of Venus Victrix   72

7.4. Theatre of Pompey: orchestra and *scaenae frons*   72

7.5. Existing state of segment of *cavea* substructure   73

7.6. Theatre of Pompey: exterior of *cavea*   73

7.7. Preliminary massing model of the Theatre of Pompey   75

8.1. Sagalassos, the primary test site of the Murale project in southern Turkey   78

8.2. Overview of the Sagalassos site   78

8.3. Views of the Roman bathhouse at Sagalassos and a model   79

8.4. Reconstruction of the stratigraphy from photographs   80

8.5. 3D CAD model of the Nymphaeum superimposed on the actual site   81

8.6. Prototype 3D camera, with image detail   81

8.7. Statue of Dionysos, with two views of the extracted 3D model   82

8.8. Sherd found at Sagalassos, with two views of the 3D reconstruction   82

8.9. The old bathhouse and surrounding landscape at Sagalassos   83

8.10. Example images of building material and synthetic textures   84

8.11. Example of texture knitting   85

8.12. An example of modern Sagalassos texture, with manual segmentation into basic ground cover types   85

8.13. Manual segmentation of the Sagalassos terrain texture shown in Figure 8.14   86

8.14. Synthetically generated Sagalassos landscape textures   86

9.1. The basic geometry of a triangulation scanner   90

9.2. A conventional flatbed scanner used at Heriot-Watt University   91

9.3. A commercial triangulation scanner, with a copy of a Roman bust of Fortuna   92

9.4. A detail from the polygonal mesh formed from the scanned head of Fortuna   92

9.5. A virtual scene   92

9.6. A schematic diagram of a conventional pulsed time-of-flight ranging system   93

9.7. A schematic diagram of the time-of-flight ranging system based on time-correlated single photon counting   94

9.8. The TOF-TCSPC optical sensing head   94

9.9. An example of raw histogram data collected from the TOF-TCSPC sensor   95

9.10. A picture and a depth image of a cast of a Roman frieze, scanned using the TOF-TCSPC technique   96

10.1. A stimulus (photograph of a torso) with superimposed ellipses   100

10.2. The result of a quarter of an hour's sampling, and the 'pictorial relief'   101

10.3. Pictorial reliefs for two observers of the same stimulus   102

10.4. Pictorial relief compared with two photographs   102

10.5. Pictorial reliefs of one observer for two different viewing conditions   102

10.6. Scatterplots for the pictorial reliefs of one observer for the same stimulus   103

10.7. Scatterplot for the pictorial reliefs of one observer for the Brancusi *Bird* sculpture   103

10.8. A multiple regression for the pictorial reliefs of one observer for the same stimulus   105

10.9. A stimulus and two pictorial reliefs obtained via two different tasks   105

11.1. Schematic diagram of coin striking   109

11.2. Idealised diagram of die links   110

11.3. Two products of the same die from a modern experimental striking   111

11.4. Coins from the same reverse punch (R91), demonstrating variability according to angle and penetration of strike (courtesy of H. S. Kim)   111

11.5. Coin struck off-centre   111

11.6. Double-struck coin, with the letters SC and wreath off-set   111

11.7. Die break across the back of a turtle   111

11.8. Ear of barley overstruck on a winged horse (Pegasus)   111

11.9. Two coins from the same die, one holed, the other slightly double-struck   112

11.10. Two coins from the same die, both cut and one very worn   112

11.11. Two countermarks, of an imperial head and a monogram, to the right of the imperial bust   112

12.1. Principal shapes of Italian *terra sigillata* occurring in Noricum   115

12.2. Poor quality of appliqué decorations due to careless handling   116

12.3. Map of the Mediterranean   116

12.4. Composition patterns of Italian *terra sigillata* with appliqué decoration   117

12.5. Different drawings of the same prototype in comparison with a photograph   117

12.6. A series of dolphins and masks as examples for the creation of prototypes   118

12.7. Principles of photogrammetry   118

12.8. Field of control points   119

12.9. Data-transformation steps to produce archaeological drawings   120

12.10. IAAD screenshot   121

12.11. Accuracy test comparing a stereo image pair   122

12.12. Manufacturing differences of appliqué decorations   122

13.1. A frontier point is the intersection of two contour generators and lies on an epipolar plane which is tangent to the surface   125

13.2. The outer epipolar tangents correspond to the two epipolar tangent planes which touch the object   125

13.3. Two discrete views showing 17 epipolar tangents in total   126

13.4. Under circular motion the image of the rotation axis l_s, the horizon l_h, and vanishing point v_x are fixed throughout the sequence   126

13.5. The circular motion will generate a web of contour generators around the object   126

13.6. Three views from circular motion provide six outer epipolar tangents to the profile   126

13.7. The motion parameters can be estimated by minimising the reprojection errors of epipolar tangents   127

13.8. Images from an uncalibrated sequence of a haniwa taken under circular motion and from arbitrary camera positions   128

13.9. Camera poses estimated from the sequence of the haniwa   128

13.10. Triangulated mesh of the haniwa model   128

13.11. Refined model of the haniwa after incorporating arbitrary general views   129

13.12. Eight images of an outdoor sculpture acquired by a hand-held camera   129

13.13. Triangulated mesh of the outdoor sculpture of a horse   130

14.1. Three stages in the reconstruction of the face of the Copenhagen mummy   131

14.2. The reconstructed head of Philip II of Macedon   132

14.3. The reconstructed head of Karen Price, and a photograph taken when she was alive   132

14.4. The reconstructed face of the Great Harwood victim, and a photograph of Sabir Kassim Kilu in life   134

14.5. Grave Circle B at Mycenae: plan showing possible kinship links   134

14.6. Reconstructed heads from Grave Circle B at Mycenae   136

14.7. The face of Gamma 55 from Mycenae   136

14.8. The face of Gamma 58 from Mycenae   136

14.9. The face of Alpha 62 from Mycenae   137

14.10. The coffin of Seianti in the British Museum   138

14.11. The skull, reconstructed face, and coffin 'portrait' of Seianti   139

14.12. Photocomparison of the reconstructed face of Seianti and her coffin 'portrait'   140

14.13. Portrait panel and reconstruction of mummy EA 74718   141

14.14. Portrait panel and reconstruction of mummy EA 74713   141

14.15. Portrait panel and reconstruction of mummy Æ IN 1425   142

15.1. The scanning of Hermione in
       progress   146

15.2. The mummy with 3D
       visualisations of the whole skeleton   146

15.3. The mummified face and
       volume-rendered skull of Hermione   146

15.4. The facial scanning system used at
       University College, London   147

15.5. Average female facial surface
       warped to the mummy skull   148

15.6. Reconstructed facial surface of
       Hermione   148

15.7. Coffin portrait, and texture
       mapping onto reconstructed face   149

15.8. Bust of Hermione milled in dense
       polyurethane   149

Table

4.1.   Location of runic inscriptions   37

Notes on Contributors

Ghassan Alusi is a Consultant Ear, Nose, and Throat Surgeon based at St Bartholomew's Hospital, London. He has a special interest in the application of computer technology to image guided surgery and has developed systems for diagnosis, simulation, and surgical guidance in ENT surgery.

Alan K. Bowman is Camden Professor of Ancient History and Fellow of Brasenose College, University of Oxford. His research interests are papyrology and the social and economic history of the Roman empire. He is the author of *Life and Letters on the Roman Frontier, Vindolanda and its People,* and *The Vindolanda Writing-Tablets* (with J. D. Thomas).

Sir Michael Brady is BP Professor of Information Engineering Science at the University of Oxford. His research interests are primarily in medical image analysis, with a strong concentration on the early detection of breast cancer and degenerative brain disease. He is the author or editor of several books, most recently *Mammographic Image Analysis,* as well as over 300 articles on Image Analysis, Robotics, and Artificial Intelligence. He is Director of the EPSRC/MRC Interdisciplinary Research Consortium on Medical Images and Signals, a Fellow of the Royal Society, Fellow of the Royal Academy of Engineering, and Founding Director of Mirada Solutions Ltd.

João Campos studied human facial profiles through automatic methods of segmentation, description, and classification for his PhD, having joined the Division of Medical Graphics and Imaging in the Medical Physics Department at University College, London, in 1987. Now at the Centre for Health Informatics and Multidisciplinary Education, UCL, he is collaborating on the development of a computer-based training tool for radiologists conducting a breast screening programme. His research interests include medical imaging and image processing techniques, and image visualisation.

Roberto Cipolla is Professor of Information Engineering at the University of Cambridge. His research interests are in computer vision and robotics, and include the recovery of 3D shape from uncalibrated images and visual tracking. He is the author of *Active Visual Inference of Surface Shape* (Springer, 1995) and co-author of *Visual Motion of Curves and Surfaces* (Cambridge University Press, 2000). He has edited five volumes and contributed numerous articles to journals and international conferences on computer vision and robotics.

Geoffrey Cross finished his PhD thesis, 'Surface reconstruction from image sequences: texture and apparent contour constraints', as part of the Visual Geometry Group at Oxford University in 2000. Since then, he has worked as a Researcher at the General Electric Research and Development Center, USA, and now works in the UK as a computer vision consultant. He has published a number of papers in the area of surface reconstruction and 3D reconstruction from image sequences.

Hugh Denard is Lecturer in Theatre Studies at the University of Warwick. He has published a number of articles on the use of virtual reality in theatre-historical research, and co-directs related projects funded by the Leverhulme Trust and Joint Information Systems Committee. He studies the reception of ancient drama in performance, and edits the online academic journal *Didaskalia*.

Ulrike Fastner is now employed in a private photogrammetric company after being an assistant in the Institute of Photogrammetry at the Technical University of Graz.

Andrew W. Fitzgibbon is a Royal Society University Research Fellow working in Oxford University's Visual Geometry Group. His research interests are in computer vision, particularly the inference of 3D information from multiple 2D images. He has received the IEEE Marr Prize, the highest in computer vision, and software based on his work has recently won an Engineering Emmy Award for significant contributions to the creation of complex visual effects. His current research interests lie in the analysis of scenes containing complex lighting and motion, such as water, glass, and fire.

Michael Greenhalgh is the Sir William Dobell Professor of Art History at the Australian National University. He is the author of books such as *Donatello and his Sources, The Survival of Roman Antiquities in Mediaeval Europe,* and *The Classical Tradition in Art.* Among others, his two main research interests are the survival and reuse of *spolia* from the ancient world, and the application of computers and the web to the teaching and learning of art history.

Helmer Gustavson is a runologist in the Runic Unit, a part of the Documentation and Research Department of the National Heritage Board in Stockholm, Sweden. He is an expert on the runic inscriptions of the Viking Age and the Middle Ages, and co-publisher of the scholarly series *The Runic Inscriptions of Sweden* (*Sverige Runinskrifter*).

Christopher J. Howgego is Senior Assistant Keeper (Roman Coins) in the Ashmolean Museum, and Reader in Greek and Roman Numismatics in the University of Oxford. He directs Roman Provincial Coinage in the Antonine Period Project (Arts and Humanities Research Board and the University of Oxford), and is author of *Ancient History from Coins, Greek Imperial Countermarks*, and *Studies in the Provincial Coinage of the Roman Empire*.

Jan J. Koenderink is Professor of Physics at the Universiteit Utrecht in the Netherlands. His interests are the study of consciousness and mind, the experimental aspects (human psychophysics of vision and active touch) as well as more formal aspects (mathematical theory of ecological optics and mental representation), with a few excursions into engineering domains (e.g. computer science).

Alf Linney is Reader in the Department of Medical Physics at University College, London, and leads a medical imaging group. His interests are largely focused on the use of computer graphics for the simulation and real-time guidance of surgical procedures. He is also an exhibiting sculptor. These interests have led to the development of methods of using computer graphics to reconstruct faces over skulls for both archaeological and forensic purposes.

Nicholas Molton is currently working on real-time camera tracking and SLAM (simultaneous localisation and mapping) in the Robotics Research Group at the University of Oxford. He has previously worked for the computer vision company 2d3 and on other vision projects at Oxford University, and gained his DPhil degree in 1995.

Richard A. H. Neave retired as Director of the Unit of Art in Medicine at the University of Manchester in 2000. Having been in the forefront of facial reconstruction in Europe, he is now Honorary Medical Artist at Manchester University, and is occupied with archaeological forensic and medico-legal work. With A. J. N. W. Prag he wrote *Making Faces Using Forensic and Archaeological Evidence* (1997; repr., 1999), and he has contributed papers on facial reconstruction and photocomparison to numerous books, including *Seianti Hanunia Tlesnasa: the Story of an Etruscan Noblewoman* (2002).

Xiao-Bo Pan was formerly Research Assistant at the Department of Engineering Science, University of Oxford, and is currently employed at Mirada Solutions, based in Oxford. She is the co-author of an article on shadow stereo, image filtering, and constraint propagation. Her research area includes image processing and stereo vision.

Marc Pollefeys is Assistant Professor of Computer Vision in the Department of Computer Science at the University of North Carolina at Chapel Hill (USA). His main interests are in computer vision, image-based modelling and rendering, image and video analysis, and multiple view geometry.

A. John N. W. Prag is Keeper of Archaeology, and Reader in Classics and Ancient History, at the Manchester Museum in the University of Manchester. As well as facial reconstruction his research interests and publications include ancient DNA, chemical analysis of Greek pottery, and Greek iconography. With R. A. H. Neave he wrote *Making Faces Using Forensic and Archaeological Evidence* (1997; repr., 1999), and most recently contributed to and co-edited (with Judith Swaddling) *Seianti Hanunia Tlesnasa: the Story of an Etruscan Noblewoman* (2002).

Marc Proesmans is Chief Technical Officer at Eyetronics, a spin-off company at the Katholieke Universiteit Leuven, specialising in 3D scanning, modelling, and animation.

Paul Robertson is Research Scientist at the Department of Computer Science and Artificial Intelligence, Massachusetts Institute of Technology, USA.

Veit Schenk was a Junior Research Fellow at Wolfson College, University of Oxford, and is currently employed at Mirada Solutions, based in Oxford. His research interests are primarily in image analysis and computer vision, with the long-term goal of developing computer vision guided autonomous systems using robust image analysis methods inspired by the human visual system.

Eleni Schindler Kaudelka is a senior staff member of the excavations on the Magdalensberg carried out by the Landesmuseum Kärnten. Her interest is centred on interdisciplinary research on Roman ceramics.

Jan O. H. Swantesson is a Senior Lecturer in Physical Geography at Karlstad University, Sweden. He is an expert on rock weathering. Current research projects deal with coastal erosion, deterioration of Bronze Age rock carvings, and the reading of runic inscription with help from micro-mapping techniques.

Melissa Terras is Lecturer in Electronic Communication at University College, London. She gained an MA in art history and English, and an MSc in IT and the humanities (both at Glasgow University), before going on to complete a DPhil in engineering science at Oxford, on the development of an intelligent system to aid papyrologists in the reading of the Vindolanda texts. Her research interests include knowledge elicitation, artificial intelligence, and image processing, as well as virtual reality, and humanities computing in general.

Roger S. O. Tomlin is University Lecturer in Late-Roman History at Oxford University, and Editor of *Roman Inscriptions of Britain*. He is the author of *Tabellae Sulis* and papers in Roman history and epigraphy.

Carlo Vandecasteele is Professor of Environmental Technology and Chairman of the Department of Chemical Engineering of the Katholieke Universiteit Leuven in Belgium. His research area is waste and waste water treatment and environmental analytical chemistry. He has a strong personal interest in ancient culture and archaeology.

Luc Van Gool is part-time Professor at the Katholieke Universiteit Leuven, where he heads the computer vision group VISICS, and part-time Professor at ETH Zurich, where he leads the BIWI image processing group. His main interests lie in 3D reconstruction, image recognition, virtual and augmented reality, and applications in cultural heritage.

Karel Van Lerberghe is Professor in Assyriology at the Katholieke Universiteit Leuven and Chairman of the Department of Oriental and Slavic Studies. He is the head of the KU Leuven missions to Tell Beydar and to Tell Tweini (Syria). He studies and publishes cuneiform texts from Tell Beydar and Tell ed-Der (Iraq) dating to the late third and early second millennium BC.

Johan Van Rompay is an electrical engineer at the Katholieke Universiteit Leuven in Belgium. He works on software implementations of algorithms for machine vision and automatic visual inspection applications.

Andrew M. Wallace is Professor of Image and Signal Processing in the School of Engineering and Physical Sciences at Heriot-Watt University, Edinburgh. He has particular interests in 3D image acquisition and analysis and in the development and implementation of parallel algorithms for image processing and computer vision.

Patrick Wambacq is Professor of Electrical Engineering at the Katholieke Universiteit Leuven in Belgium. His research area was image processing targeted at machine vision applications, but recently he switched to speech processing, with an emphasis on speech recognition.

Kenneth Wong completed his PhD ('Structure and motion of surfaces') at Cambridge, and is now an Assistant Professor at the University of Hong Kong.

Alexey Zalesny is a Postdoctoral Researcher at the Computer Vision Laboratory (BIWI) of ETH Zurich. His field of work is the analysis, synthesis, and segmentation of texture.

Andrew Zisserman is Professor of Engineering Science at the University of Oxford, and heads the Visual Geometry Group. His research interests include geometry and recognition in computer vision. He has authored over 100 papers and is author/editor of eight books. He has twice been awarded the IEEE Marr Prize, the highest in computer vision, as well as the Engineering Emmy Award, television's highest technical honour.

Preface

The papers that make up this volume are based on the lectures given at a joint British Academy/Royal Society discussion meeting entitled Images and Artefacts of the Ancient World, which was held at the Royal Society in London in December 2000. The meeting was attended by approximately 100 people, whose interests spanned a remarkable diversity of disciplines — from ancient history to electronics — but who were united in their interest in applying recent developments in imaging, image analysis, and image display and diffusion to objects of material culture in order to enhance historians' understanding of the period from which the objects came (in all cases here treated, the remote past, but not only the classical world). A further consideration, which we specifically addressed in this forum, is the fact that such imaging techniques now offer the researcher valuable insurance against the processes of deterioration to which such artefacts are inevitably subject.

The idea for such a discussion meeting was an outcome of our joint research project, generously supported by the Leverhulme Trust in 1997 and by the Engineering and Physical Sciences Research Council (EPSRC) in 1998–2001, on the image enhancement of incised writing-tablets from Roman Britain, in which we are attempting to develop new signal-processing techniques for damaged and abraded documents with three-dimensional writing and, in many cases, palimpsest texts. Two papers arising directly from this ongoing project are included as the first two chapters in the present volume.

Broadly speaking, we hoped that the meeting, and the subsequent volume, would provide a forum for wide-ranging discussion of state-of-the-art computer-based imaging. To this end, we invited sixteen presentations, of which half were technical and scientific and half archaeologically or historically focused. To emphasise further these two perspectives, each session comprised both technical and historical papers. The intention was to stimulate further research and to encourage people to talk introspectively about their methods and problems, to analyse their own processes of problem-solving and visual cognition, and, in particular, to focus in a very precise and realistic way on what is empirically desirable and technically feasible. The timing of the meeting was appropriate because it seemed to us — and we contend that the success of the meeting supports this view — that there was a very great deal of productive research in progress, but perhaps not yet too many genuinely innovative and well-founded projects, to make this relatively small group of speakers unrepresentative or meaningless. We had initially felt that the programme format allowed relatively little time for discussion and we were worried that participants might feel that they had not had enough time to discuss all the issues which arose. In the event we were able to extend the sessions at the end of each day and to provide some extra discussion time. The speakers were also very punctilious in keeping to the time limits and, in the event, we did not feel seriously short of discussion time.

The organisational structure was from our point of view excellent. The meeting was held in the premises of the Royal Society but was jointly funded by the Society and the British Academy at a level which did not constrain the programme or the roster of participants. We are grateful to both institutions for their generous support. At the Academy, Rosemary Lambeth provided essential help in setting up the programme. We would like to express our gratitude in particular to Froniga Lambert of the Royal Society. Her assistance at all stages of the planning and during the meeting itself was invaluable and supplied unstintingly and cheerfully. The attitude of all the staff at the Society helped make the atmosphere extremely relaxed and pleasant and this contributed in considerable measure to the intellectual benefit derived from the meeting. We were also very appreciative of the technical assistance with the audio-visual aids required.

We are conscious of the fact that the publication of the papers given at the meeting has involved more delays than we would ideally have wished and we regret that we have been unable to include a published version of one of the papers. Despite the delay we believe that the papers still offer valuable discussion of many aspects of state-of-the-art computer-based imaging which will provide some benchmarks for future research in this area. We are particularly pleased that the Society and the Academy were willing to undertake joint publication of the monograph and we are very grateful to

James Rivington and his staff at the British Academy, who have patiently steered the volume through to publication with unfailing helpfulness and courtesy. We are also grateful to Charles Crowther, Melissa Terras, Margaret Sasanow, and Livia Capponi for help at various stages in the preparation of the manuscript.

AKB
MB
Oxford, January 2003

Introduction

Michael Brady & Alan K. Bowman

Background

The genesis of the papers in this volume was a joint research project on the image enhancement of ancient documents, directed by the present editors, involving the Centre for the Study of Ancient Documents and the Robotics Research Group (Department of Engineering Science) at the University of Oxford. In the past few years both the engineers and historians working on this project have become aware of a great deal of research on image enhancement and processing, applied to a variety of different scientific and humanistic fields, and we felt that it would be timely and profitable to bring some of these researchers and projects together in a forum where techniques and results could be discussed in a way which would maximise the possibilities of cross-fertilisation, of generalising techniques from one category of object or problem to another, and encourage people to think laterally and in a genuinely interdisciplinary way about advancing the techniques of imaging and the visual understanding of the objects under study. It hardly needs to be said that we were not the first to address these problems[1] and to attempt a fruitful interdisciplinary dialogue, but it seemed to us that there was a great deal of potential interest and profit in a discussion focused in particular on problems, methods, and processes of problem-solving and visual cognition. A significant further dimension is the improvement of techniques of providing high-quality images of important and valuable collections of original artefacts to scholars who cannot always study the originals directly, although in almost all cases direct autopsy of originals is preferable.[2]

It should be stressed from the outset that our own work with fragmentary and damaged documents inevitably led us to the position that we would not suggest that the analysis of images — with or without the aid of special light sources and cameras which transcend the exquisite abilities of the human eye — is simply a matter of physics, engineering, or uninformed transcription. Indeed, it is apparent that the historian[3] acts out a perceptual-cognitive task of transforming often noisy and impoverished signals into semantically rich symbols

that have to be set within a cultural and historical context. We set out explicitly to undermine the simplistic notion (perhaps now encountered less frequently than a decade ago) that the engineering scientist can offer the historian a range of off-the-shelf techniques of which one or another will provide an automated solution to problems in 'reading' and interpreting visual data. The historian's interpretative skills are very often directed to evidence which is imperfect or fragmentary, are always subjective, and always rely on a body of knowledge and expertise acquired from experience with similar data. The theme of perception and cognition was one which was repeated throughout the meeting, both in the presentations and in the lively discussion which ensued, and is explicit in several of the papers, in particular those by Brady et al., Bowman and Tomlin, and Koenderink.

Conversely, we have also become aware that engineering scientists, equipped with a range of sophisticated techniques, equipment, and highly specialised knowledge, are not always as aware as they might be of the range and the exact nature of problems faced by historians in interpreting objects of material culture. No doubt this is largely because researchers in different disciplines do not often have the opportunity to explain to each other in broadly comprehensible terms precisely what they are doing and how. In providing such an opportunity, it is our hope that readers of this volume will be aware of the extent to which these scientific contributors, at least, are thinking about subjectivity of interpretation, visual cognition, and the need to improve methods of presenting evidence in ways which feed directly back into their own scientific thinking and encourage genuine innovation in their approach to developing methods of image enhancement. Our impression at the end of the meeting, and at the end of reading and editing the papers in this volume, is that this hope will be realised to a significant extent.

The Papers

Our grouping of the papers in this volume largely follows that of the meeting itself, but we emphasise that there is inevitably some overlap between the subjects and methods of the papers in the various sections, and that the techniques and the problems addressed under the different section headings return, explicitly or implicitly, to the common themes of the discussion, outlined above.

These fifteen papers offer a good range of treatments of artefacts and of techniques, many of which

---

[1] See, for example, Higgins et al. (1996).

[2] This is the central concern of the paper by Greenhalgh (Chapter 5). As regards autopsy, significant exceptions are cases where special techniques such as infra-red photography or digitisation reveal features not visible to the naked eye.

[3] We use the term 'historian' only as a convenient shorthand. The disciplines and subject areas covered by our contributors include archaeology and its constituent specialisations, as well as art and architectural history, papyrology, and palaeography.

could develop applications to a broader range of material objects.[4] Inevitably, the coverage is not comprehensive. Some scientific techniques (e.g. confocal microscopy) are at present difficult to apply from a practical point of view (expense, portability, and so on), but advances in the development of imaging equipment will certainly continue to make a broader range of techniques more practical and affordable, as to some extent they already have. We could not hope to include artefacts of material culture in all categories. And we emphasise that all the artefacts which we have chosen to include require techniques of imaging in three dimensions. This field of enquiry seems to us to offer the most challenging problems and the most potentially rewarding results.[5]

Imaging Documents

Written documents form a significant proportion of the artefacts that inform our understanding of the ancient world. Many texts were written on whatever was the most convenient and enduring material locally or easily available: stone, metal, papyrus, potsherd, wood. A great many of these, including many of the best-known collections of very early written material, were incised or inscribed. Although large numbers of such texts have been published, it should be emphasised that there are still many writing-tablets of various kinds lying unread and unpublished in museum collections, and that corpora of published texts are, or should be, continuously subjected to critical evaluation, and re-reading and improvement, for which the availability of better images is vital. Vandecasteele and his colleagues (Chapter 3) discuss the imaging of cuneiform tablets, tablets of moist clay on which an 'intellectual' (or a literate official) could inscribe a message with a stilus before baking the tablet. Treatment with an ammonium aerosol often yields significant improvement in the visibility of the writing, which is then digitised by imaging under low-angle raking illumination. These digital techniques are now being used continuously by the authors, in the field, on archaeological sites in the Middle East.

Brady, Bowman, and their colleagues describe work on a different type of inscribed documents: wooden stilus tablets, written during the Roman occupation of Britain and originating mostly from Vindolanda, a Roman fort lying just south of Hadrian's Wall. Brady and his colleagues (Chapter 2) describe the appearance of stilus tablets and then outline a novel method called shadow stereo for picking out the written strokes. Towards the end of the article they illustrate recent work

that combines the image analysis that is the main focus of the article with an artificial intelligence programme to offer an interpretation of a line of Latin text. Interpretation is the subject of the article by Bowman and Tomlin (Chapter 1), who show how improved readings of a text developed over time, and suggest that the process of interpreting a text involves mobilising information of different sorts, about epigraphy, the Latin language, and the historical context in which the text was produced, as well as knowledge about what the message might mean.

Laser Imaging

The range of options for imaging, particularly three-dimensional artefacts, has increased dramatically with the invention of the laser and with the development of electronic devices that are fast, compact, and cheap. A laser emits light that is coherent and which is, for many purposes, monochromatic — essentially, only a single wavelength is present, unlike the broad spectrum in normal 'white' light. Laser light can be focused very precisely, and, in combination with special transmit-and-receive devices, can measure distance to the imaged object precisely by measuring the 'time of flight'. The problem is that light 'flies' rather quickly, at approximately 2.29×10^9 metres per second. Estimating depth variations that are measured in millimetres poses a massive problem for instrument design.

Wallace (Chapter 9) reviews the two principal techniques in current use for 3D active imaging using projected laser light, then describes a technique to solve the problems of occlusion of viewpoint and multiple reflection. His paper outlines a new type of time-of-flight sensor which he has developed, and explains how it promises an improved specification when compared to existing sensors, and opens up new applications, for example scanning of transparent objects.[6] The system described by Wallace is currently confined to the bench,[7] so it is only of academic interest to historians who wish to study artefacts in the field, such as rune inscriptions, which cannot be brought into the laboratory. In the meantime, there are other ways of going about acquiring such images. To this end, Swantesson and Gustavson (Chapter 4) describe a portable micro-mapping device for measurements of the micro-relief of rock surfaces in the field. The central unit of the machinery is a commercially available laser gauge probe that can obtain height values with a resolution of 0.025 mm within a measurement range of 100 mm. The mechanism comprises motors to drive the sensor over the rock surface and a computer for control, data storage, and data analysis. The micro-mapping device has also been used for

---

[4] Our collaborative project on the stilus tablets has suggested further research on techniques of recovering damaged texts inscribed on stone by investigating sub-surface damage caused by the impact of the stonecutter's chisel. This research will be supported by the University of Oxford's Research Development Fund in 2003–5.

[5] Techniques of infra-red imaging of documents written in carbon-based ink (e.g. papyrus texts, ink writing-tablets, Dead Sea Scrolls) are relatively well understood and serviceable: see Bowman et al. (1997), 169–76.

[6] At the meeting, Professor Fred Fitzke (University College, London) gave a compelling lecture on the potential of various forms of (affordable) microscopy, in particular confocal scanning optical microscopy. Unfortunately, a written version of his presentation was not available for inclusion in this volume.

[7] As is that of Fitzke (see previous note).

detailed documentation of about 10 runic inscriptions in Sweden and Norway.

3D Reconstruction of Artefacts

Is it possible to build inside a computer an accurate three-dimensional model of a complex three-dimensional shape such as a sculpture? We know that the human vision system uses the slight differences between the images arriving at the two eyes to gain some appreciation of depth, though that is not the only depth perception system available to humans (as Koenderink explains). In human stereo vision, the eyes are close together, and so, for objects that are more than a few centimetres away, they see approximately the same part of an object's surface. Photogrammetry comprises a set of techniques for combining a few images to form a 3D reconstruction of an object. This technique is the basis of the article by Schindler Kaudelka and Fastner on the appliqué decoration of the common type of ancient pottery known as Italian-style *terra sigillata* (Chapter 12). The questions which they address include: is it possible to trace a generation line from the prototype of a die to the respective appliqué decoration on any vessel? Is it possible to establish a basic pattern for a chronological approach? Is it possible to define a typological set of dies for single potters? The importance of these issues for classical archaeologists and historians cannot be overstated, particularly because reliable methods of chronological sequencing and dating of pottery underpin a huge number of archaeological site analyses, upon which in turn rest hypotheses of historical development of much broader significance.

Stereovision and photogrammetry are typically — though not always, as these authors show — limited to a small number of views taken from nearby vantage points. This makes these techniques particularly useful for '2½D' objects, which are essentially curved surfaces. Reconstructing an accurate 3D model of an object such as a sculpture of a horse requires that views be combined from a much greater set of vantage points. Imagine that the sculpture were placed at the centre of a large sphere from every point of which it could be viewed, along the surface normal at that point. This is known as the *view sphere*. The papers by Fitzgibbon and colleagues (Chapter 6) and by Cipolla and Wong (Chapter 13) show how, using the silhouettes of an object, viewed from a large number of points on the view sphere, accurate reconstructions can be made.

Reconstruction of Buildings and Spaces

The papers by Denard, Van Gool and his colleagues, and Greenhalgh (Chapters 7, 8, and 5, respectively) increase the spatial scale of the artefact studied to whole buildings and even to entire archaeological excavation sites. These studies raise important issues about the scale of the individual objects to be imaged, and about the most effective means of delivering large quantities of images in a manageable format for research and teaching. Denard notes that despite its great historical, architectural, and cultural importance, until now there has been no modern scientific survey of the relatively exiguous remains of the Theatrum Pompeianum, one of the most important buildings of late Republican Rome. He describes an ambitious project to use virtual reality technologies to visualise the theatre, and, equally importantly, to test alternative hypotheses about the structure. Three-dimensional computer modelling will be widely used.

Van Gool and his colleagues describe initial progress on a European archaeological project, Murale, which aims to use state-of-the-art computer vision techniques to study the ancient city of Sagalassos, in modern Turkey, which was destroyed by an earthquake in the sixth century AD. The Murale project aims, quite generically, to develop tools for the analysis, archiving, and visualisation of archaeological sites and their artefacts. These include systems to model the finds in 3D (from small scale to large scale), tools to virtually restore and complete finds, and tools to archive and retrieve information about the finds, with access over the Internet.

Greenhalgh's paper describes a website (ArtServe) which already contains over 160,000 examples of architecture and works of art, and which is primarily intended for teaching. The website has been designed to give students a heightened sense of reality, enabling them to navigate through a building (or painting) and to link to appropriate contextual information.

Depth Perception from Relief

The papers grouped under this heading tackle a very difficult area of human vision and perception, one which is not fully described or understood and which offers huge potential in attempts to replicate the human visual processes in computer-based imaging. Howgego's paper (Chapter 11) offers a rather different perspective from the others in the volume, since he does not describe work in progress, but discusses the potential contributions which image analysis might make to numismatics. He argues that the methodology of the die study and its application from the 1870s for mint attribution were also linked to a major revolution in imaging, namely the advent of photography. However, he also notes that die studies are of limited use for attribution and quantification. The ability to compare images quantitatively would overcome many of these limitations and bring about a 'quantum leap' in numismatics. This seems to us an area in which the potential can be realised by a very sensitive and penetrative analysis of the ways in which existing techniques can be tuned in relation to a set of very specific requirements and limitations which differ in some significant respects from those presented by other types of artefacts discussed here.

Koenderink is concerned with a perceptual process of fundamental importance, which relates in one way or another to almost all the specific categories of artefacts under consideration (Chapter 10). When one looks at a single shaded photograph of a three-dimensional object, one often has a compelling sense of the 3D structure. However, when different observers' estimates of that 3D structure are tested experimentally, they vary considerably — from each other and (by implication) from the 'ground truth'. Koenderink demonstrates a technique for testing such estimates and then suggests possible metrics for exploring 'pictorial space'. He notes that 'the theory of the structure of pictorial space is closely related to a general understanding of the intrinsic ambiguity of pictures and to an understanding of observers as efficacious agents'. We note not only that many objects have traditionally been studied in two-dimensional photographic representations (and that for a significant universe of lost or irretrievably damaged objects photographs are now the only representations available), but also the obvious and paradoxical feature of this type of research: improvements in three-dimensional imaging, are of course, presented literally in only two dimensions to the student or researcher looking at them on a computer screen.

Reconstruction of Faces

The volume concludes with two papers that address the reconstruction of faces, certainly comprising the most distinctive physical characteristics of the individual. For this reason, face reconstruction is a problem that is of as much interest to those reconstructing mummified faces and damaged sculptures as it is to those undertaking reconstructive surgery and criminal investigations. Following the practice of a medical artist who primarily relies on hand-and-eye techniques of illustration, Neave and Prag outline the steps to reconstruction of a face from an estimation of the skull shape (Chapter 14). This involves building up facial muscles using statistics for flesh thickness, and then adding a layer of clay for the subcutaneous tissue and skin. They stress that it is the proportions of the underlying armature — the skull — that dictate the form of the building, or the face, and give it its individuality. Several case studies are described in detail, including the evidence from the tomb of Philip of Macedon, father of Alexander the Great. As in the case of reading documents, issues of intuition, individual expertise, and subjectivity are emphasised here. We are appositely reminded that their machine is not a factotum and that the methods of research discussed here are not aimed at replacing human skills but at complementing them and making them more effective.

How this can be done in this area of research is demonstrated by Linney and his collaborators, who show how accurate, detailed imaging and the use of three-dimensional reconstructed models now play an important part in planning surgical procedures. They describe part of a collaboration with the British Museum,

in which they have used computer graphics techniques designed to simulate facial surgery to reconstruct the 3D facial image of a mummified face from X-ray computer tomography (CT) scans, and have used the mummy portrait to texture-map a face onto the 3D reconstruction.

The Outcome

It is often the case that meetings leave a warm glow of appreciation that soon fades. The nature of the venture means that it is too early to say whether the meeting has made a difference, and how, except in so far as it has already stimulated conversations between researchers who had not conversed before (not, however, a trivial result). We hope and believe, however, that, beyond this already important point, the meeting and the volume will have lasting value to historians and image scientists alike. Indeed, we think that the volume will set a standard and a guideline for interdisciplinary research and for exploration of further possible applications of imaging techniques to a wide range of artefacts from the ancient world (and by extension to other areas also).

What are some of the outcomes of the meeting, apart from forging a nascent community? First, to return to the first paragraph of this Introduction, there is a recognition of the need to use image analysis as a basis for studying the process by which the historian effects an interpretation of an artefact. Ambiguity is omnipresent, as stressed by Bowman and by Koenderink. Collaborative projects should involve someone familiar with the computational techniques for representing and mobilising knowledge of many different kinds — that is, artificial intelligence. There is much to be explained and explored in this area, discounting any purely simplistic notion that the machine is capable of *creating* interpretative techniques.[8]

Second, a wealth of new technologies and techniques were discussed that have enormous potential for the historian. Fast, accurate ranging technologies such as the single photon counting technology described by Wallace will surely reduce sharply in price and size, perhaps even to the point that they become instruments that are usable in the field. The 3D reconstruction techniques described by Van Gool, Fitzgibbon, Cipolla, Linney, and Brady are of immediate applicability to a range of problems described by the historians. Denard makes a powerful case for the way in which virtual reality can support theory formation and testing by the historian.

Third, just as the scientists have a number of novel technologies to offer the historian, the historians posed a wealth of challenges that will keep image scientists busy for many years to come. For none of the many kinds of artefacts described in the volume can it be said that their analysis is remotely 'solved'. Nor will this ever happen in any definitive way. Solutions to one problem or to one set of problems will merely lead us on to the

---

[8] In our own case, we have been fortunate to have available the skills of Dr Paul Robertson and Dr Melissa Terras.

next, more challenging set and will force us to reformulate the questions which we can put to the material and to the machines which process it. The pioneering computational experiments described in the various papers provide some encouragement but, in every case, there is a long, long way to go. We already have some novel problems posed, but as yet unaddressed, let alone answered. Evidently, there is much to do. The present volume is just a very small, but we hope significant, beginning.

References

BOWMAN, A. K., BRADY, J. M. and TOMLIN, R. S. O. (1997), 'Imaging incised documents', in A. K. Bowman and M. Deegan (eds.), *Imaging Documents, Literary and Linguistic Computing* 12/3: 169–76.

HIGGINS, A., MAIN, P. and LANG, J., eds. (1996), *Imaging the Past*, British Museum Occasional Paper 114 (London).

CHAPTER 1

Wooden Stilus Tablets from Roman Britain

Alan K. Bowman & Roger S. O. Tomlin[1]

I. The Nature of the Problem

The imaging of ancient documents presents a variety of challenges whose nature is determined principally by the character of the text, the material on which it is written, and the state of preservation.[2] The authors of this paper, and Professor David Thomas (of the University of Durham), have been struggling to read and to interpret difficult Latin manuscripts from Roman Britain for the past quarter-century or more. These come mainly in three forms: texts written in ink on thin wooden leaves, texts incised with a metal stilus on wax-coated wooden stilus tablets, and texts (usually 'curses') inscribed on sheets of lead.[3] It is the second category that concerns us in this paper, presenting, as it does, extremely challenging problems of imaging and signal processing.

There are at least two separately identifiable but linked sources of difficulty. The first and primary source is the problem of seeing and identifying, in abraded and damaged documents, what it is that we are trying to read, and it is this problem which primarily attracts the attention of the signal-processing expert. The other category of problems arises from what may be called the character of the text itself — those features which determine the ability of the reader to decipher and interpret it, assuming that we can actually see it as well as we need to. Although we can in principle separate these two aspects so that we would, in an ideal world, first achieve a perfect visual image and then read it correctly, in practice our ability to read even the 'perfect visual image' is less than perfect; it goes without saying that the likelihood of our achieving a correct reading diminishes the poorer the visual image is.

The difficulties can be well illustrated by the editorial history of a fragmentary text written in ink and discovered in 1974 among the writing tablets from

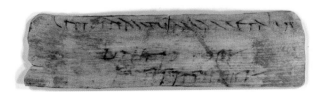

Vindolanda.[4] The text was originally read from an excellent infra-red photograph, made in 1975 by Alison Rutherford, in which the script and the individual letters in the top line were relatively clear (i.e. in itself it presented what seemed to be quite clear visual signals).[5] Despite this, a comparison of the 1983 and the 1994 transcripts shows that the best reading that we could achieve the first time round was woefully inadequate (Figure 1.1):

1983 transcript:[6] **c...io inmatura ad metalla**

Here, each letter as we transcribed it was a defensible reading and it was possible to reconstruct some known Latin words from it, if not any syntax or sense: *inmatura* would be a form of the adjective *immaturus* and the last two words could be understand as a possible reference to the known practice of condemning criminals to hard labour in the mines (*metalla*).

1994 transcript:[7] **Lepidinam tuam a me saluta**

Eleven years later we were confident that we had the correct reading: 'Greet your wife Lepidina from me'. What made the difference here was an accumulation of contextual knowledge from other tablets, discovered and deciphered between 1985 and 1994, which gave us plenty of evidence for the existence of the woman called Lepidina, the wife of the commanding officer of the unit stationed at Vindolanda between *c.* AD 97 and 104, and of the formulae commonly used at the end of personal letters of this sort. The contextual information makes a crucial difference because in Latin cursive handwriting of this general type (known as Old Roman Cursive, and in vogue until the third century AD) the letter forms themselves are, in almost all cases, far from being completely unambiguous: for example, *a* and *r* are frequently made

---

[1] The two authors have collaborated closely in all aspects of the work discussed here. Responsibility for the presentation of Part I rests primarily with AKB, for Part III with RSOT; Part II is a joint effort.

[2] For surveys of different types of material see Bowman and Deegan (1997), in particular the articles by Bagnall, Gagos, Crowther, and Bowman, Brady, and Tomlin. A range of material is available at http://www.csad.ox.ac.uk.

[3] See Bowman and Thomas (1983, 1994), and Tomlin (1988, 1998). These are far from being the only categories of written material from Roman Britain. This particular aspect of our imaging work does not address the problems presented by monumental inscriptions on stone, or by the graffiti and inscribed personal belongings classified as *instrumentum domesticum*.

[4] *Tab. Vindol.* I 25 = II 247. Readings by A. K. Bowman and J. D. Thomas.

[5] Figure 1.1 is not the original photograph but a digital scan, made by Dr John Pearce in 2000, which offers a marginally better image.

[6] Bowman and Thomas (1983), no. 25.

[7] Bowman and Thomas (1994), no. 247.

similarly; *c*, *p*, *t*, and *i* are often difficult to tell apart; and these are by no means the only potential sources of confusion.[8]

Such confusions are only too easy to illustrate. One lead tablet was first transcribed a century ago, when it was interpreted as a letter written by one Christian to another in fourth-century Britain. To the first author of the present paper, as one generally familiar with the range of documentation from the Roman empire, the text as originally read looked very suspect but he was not able to explain it, except by vaguely wondering whether it might be a modern forgery. The second author achieved a better reading and interpretation by rotating the tablet through 180°: its Edwardian editor had read it upside down, and the new reading is self-evidently correct, re-locating it to what is now a well-known genre of Latin texts.[9] This reading experience has recently occurred in the Vindolanda tablets in a still more complex and enigmatic form: one of the ink tablets discovered in the 1990s contains a text which has two lines inserted upside down in relation to the rest of the writing on the leaf.[10]

A major part of the problem, particularly in the early stages of our work on this material, was simply its novelty. It was not completely unparalleled in texts from other parts of the Roman empire (particularly Egypt), but we lacked a large existing corpus of similar texts which would give us a broad enough basis of knowledge from which we could deduce or predict with a high level of confidence what we were likely to have in any new example. Papyrologists in general proceed on the principle that what they have in a new text should ideally be able to be paralleled in an existing text, and only when they have searched exhaustively and failed to find parallels do they feel comfortable about proposing a reading or interpretation that is completely novel. As the corpus of existing material accumulates, the possible parallels increase, and the decipherment and interpretation of new texts improve accordingly. The number of surviving Latin texts from Roman Britain (let alone the empire as a whole) as a proportion of the total ever written (or even of those produced in a few decades) must be minute.[11]

Even with the advantage of a steadily expanding knowledge base, collaboration on the Vindolanda tablets makes it painfully obvious that in many cases two supposed experts working independently can come up with completely different transcriptions of the same visual signals (for which there has to be an objectively correct interpretation, i.e. what was originally written), and that they must resolve these differences by negotiating their way towards an agreed version which approximates as closely as possible to the objectively correct transcription and interpretation.

In considering how computer-based image enhancement can help the expert reader, it seems obvious that there can in principle be no simple automated process which will solve problems that arise from a combination of damaged or abraded and inevitably imperfect human cognitive and knowledge-based skills. Our curiosity about developing an effective process that can help might profitably have been directed to the fairly substantial body of ink texts like the one described above. There are, however, already reasonably effective imaging techniques in existence for digitising and enhancing ink-written texts,[12] and in fact the wooden stilus tablets — of which more than two hundred have been discovered at Vindolanda in excavations conducted since 1974, and many more lie unread in museums and older collections — present far more difficult and challenging problems, not least because they present them in three rather than two dimensions.

The cognitive and knowledge-based problems of the sort described above are in general common to ink texts and inscribed texts. Understanding them requires some introspection and analysis of the process of reading damaged, abraded, or degraded texts, which we take to be different from reading printed text on a page. First we identify the shapes of letter forms, which fall into a range of types and different hands; then we read individual letters and combinations of letters to the point where we can construct words, phrases, sentences, and finally whole texts.[13] But of course we do not normally transcribe letter by letter in a completely neutral and automaton-like fashion, and then realise that we have transcribed a word or sentence; there is a point, very quickly or perhaps immediately reached, at which we bring into play our corpus of acquired linguistic, palaeographical, and historical information which in effect predisposes or even forces us to predict how we will identify, restore, or articulate letters or groups of letters. And this is a recursive process. We do the easy bits, then make hypotheses about the problematic bits and test them in the context which we think we have established by what we can read. The process of reading papyri was well described, even if somewhat intuitively and impressionistically, by Herbert Youtie, probably the greatest papyrologist of the twentieth century, a generation or so ago and well before the existence of computers and imaging techniques:[14]

> In the face of such discouragements [sc. the poor physical state of the papyrus], in whatever combination they may occur, the transcriber repeatedly finds that his most strenuous efforts to obtain a reading are frustrated. His only hope lies in supplementing his knowledge of handwriting with as full an understanding as he can get of the scribe's purpose in writing the text. He tries

---

[8] Bowman and Thomas (1983), 51–71; (1994), 47–61.

[9] Tomlin (1994).

[10] Bowman and Thomas (2003), no. 649. Not recognised by A. R. Birley and R. E. Birley (1994), 431, no. 1, although they did note that there seemed to be an insertion in the text.

[11] Fink (1971), p. 242, noted that in the first 300 years of the Roman empire at least 225 million individual pay records would have been written for legionary soldiers, but that only two examples, each relating to a handful of soldiers, actually survive. One more has come to light since 1971.

[12] For some references see Obbink (1997).

[13] For tabulation of letter forms see Bowman and Thomas (1983), 54; (1994), 53; Tomlin (1988), 91–4. It should be noted that the exemplars are based on a relatively small sample of texts.

[14] Youtie (1973), 26–7.

to take account of the text as a communication, as a message, as a linguistic pattern of meaning. He forms a concept of the writer's intention and uses this to aid him in transcription. As his decipherment progresses, the amount of text that he has available for judging the writer's intention increases, and as this increases he may be forced to revise his idea of the meaning or direction of the entire text, and as the meaning changes for him, he may revise his reading of portions of the text which he had previously thought to be well read. And so he constantly oscillates between the written text and his mental picture of its meaning, altering his view of one or both as his expanding knowledge of them seems to make necessary. Only when they at last cover each other is he able to feel that he has solved his problem. The tension between the script and its content is then relaxed: the two have become one.

To bring these skills into the purview of 21st-century technology, which we should surely try to do, we need to begin by analysing the process again and identifying the capabilities and the limitations of this technology.

Self-evidently, all the processes of cognition involved in reading a text cannot be replicated by a machine, but some of the processes can be made easier or brought into relationship with one another more effectively. Making alphabets for individual texts, which is a standard procedure followed by many palaeographers, is an act partly of vision, partly of cognition. In Latin cursive texts, a computer-based technique of pattern recognition for letters might tell us that a certain combination of strokes is most likely to belong to *t*, *p*, or *c* (a similar group) or *a* and *r*, but it takes a combination of palaeographical and linguistic knowledge and expertise to perform the act of judgement which privileges one possibility over the others, when letter forms vary even in individual texts, as they tend to do frequently enough to cause confusion. Faced with a group of letters which could equally well be read as *par* or *tra*, the decision as to which is more likely cannot be made purely on visual criteria but requires the deployment of other skills and knowledge. And it might be argued that a palaeographer could deploy them just as well without a computer, especially in a subject where the acquired knowledge is relatively limited. We believe, however, that this is too optimistic, and that much is to be gained by the systematic classification and presentation of comparative data, both visual and knowledge-based.[15]

The problems of visibility, however, are a different matter and here we think that signal processing really can help: once again, not in producing any kind of a fully automated 'system', but in producing techniques which enable us to see better the material text on which we are trying to perform the cognitive act of 'reading', bearing in mind the ways in which we can ourselves deliberately manipulate the physical act of vision by varying angle and focal length. The stilus tablets offer an ideal corpus of material. Some few can be read wholly or partly without any artificial aid, save simple lighting; many others really do seem impossibly hard; others lie between these extremes. If we can read some of them, or only some bits of some of them, we ought to be able to use computer techniques to push forward the boundaries of legibility.

The stilus tablet is a thin, rectangular slab of wood, about the size of a postcard, with a hollowed-out centre which is filled with wax. The writing is incised with the sharp point of a metal stilus on the wax surface. The wax can be re-smoothed with the spatulate end of the stilus, obliterating the old text and allowing a new one to be written, or a new wax surface can be put on.[16] In the great majority of cases the archaeological contexts in which these objects are found has caused the wax to perish. We are thus left only with the scratches made by the stilus where it penetrated the wax to the wood beneath.

Our starting point was the notion that we could describe and analyse the visual process and attempt to replicate it more efficiently by using a computer-based signal processing technique. The attempt to read these things in the old-fashioned, manual way can be based on photographs, which offer a static image, or on the originals. Achieving optimal visibility involves adjusting the angle of lighting (which would obviously require more than one photograph if one is not working from the original), varying the angle of vision so that different parts of the incision can be seen, and using other indicators such as alignment, depth of incision, the crossing of superimposed strokes. In fact, we observed that when we do try to read the originals we tend to use lighting from different angles and to tip and tilt the surface in relation to the eye, so that the text more or less 'moves' for us. This intuitive behaviour, when he 'spotted' it, prompted the way in which Brady came to our aid.[17] What we would achieve, if only the human eye and brain were perfect, would be a series of optimal images of different bits of the text or, in actual fact, of different bits of the same strokes as they emerge more clearly under different lighting conditions and different angles of vision. But one of the respects in which the eye and brain are not perfect, as we conceive it, is that even where achieving such a series of optimal images is possible, it is impossible for the human eye to achieve a complete synthesis of them. This is where it seems that computer-based signal processing ought to be able to help. The examples which follow illustrate a variety of problems with different degrees of difficulty, the methods used, and the results we have obtained so far.

II. Two Sample Texts from Vindolanda

Tablet 797 (Figure 1.2)

This tablet is the most straightforward of the three presented here. It contains a single text written in two

---

[15] For an outline see Terras (2000).

[16] Lalou (1992).

[17] We quote from the article 'Wax tablet reveals secrets of Roman life' in the Oxford *Blueprint* for April 2001, which in effect describes how the Professor of Information Engineering went bird-watching among the ancient historians.

Figure 1.2 Vindolanda stilus tablet 797.

(a)

(b)

columns on either side of a central recessed strip.[18] We have succeeded in reading the last two(?) lines of column 1 on the left, and the five lines of column 2 on the right. We were able to do this on the basis of a small number of unprocessed scans, made with lighting from different angles. The tablet may therefore be placed at the relatively unproblematic end of the spectrum. The text itself is a letter, or rather the end of a letter, in which the writer offers a guarded apology to the recipient:

Column 1	Column 2
traces	**nunc quid**
traces	**mi irasce**
traces	**ris aliqua**
me irasce	**ndo mir**
ris	**[o]r vale**

'...angry with me. [...] sometimes I am surprised at why you are now angry with me. Farewell.'

Tablet 836

This example gives a unique insight into the correspondence between the original state of the stilus tablet with its wax coating, and the more usual state in which tablets are found, without the wax, because this was the only case in which the wax coating survived the archaeological conditions (although it dissolved during the conservation process). It therefore offers a unique opportunity for illustrating and testing the imaging techniques which are being developed with the use of variable directional lighting and phase congruency.

Fortunately the tablet was photographed with the wax writing surface more or less intact, at two different scales, and we thus have two images of a single incised text on a relatively 'noise'-free surface (Figure 1.3). When the wax was accidentally dissolved, it left the bare wooden surface with the incisions made by the stilus where it penetrated the wax. It can be seen that, as in so many tablets, there are incisions not just from the text which we could see on the surviving wax, but from at least one other text as well. We have photographs and digital scans of the tablet in this state, taken with directional illumination.

By using the images of the tablet with the wax coating we have been able to produce coherent readings of

Figure 1.3 (a) Vindolanda stilus tablet 836, before the wax dissolved. (b) Tablet 836 after the wax dissolved.

parts of the latest text written on this tablet. It should be emphasised that this text, the one on the wax surface, is relatively visible, but even so there are points where the readings of the two authors of this paper differ, and there is one area which offers particular difficulties and ambiguities. In the text presented below, letters printed in **boldface** are those which can be read with confidence; letters printed in ordinary type are read with some measure of conjecture; underlinings indicate traces of letters which cannot be identified with confidence:

binus bello suo salutem
[traces only]
]acc__**erunt in uecturas**
de_arios octo **reliquos** solues
rios nouem qua__r_r___   5
]**sam dari debeb__**
[interlinear addition?]
]**em libris**
]**dus uale**

'Albinus to his Bellus greetings . . . they have received for transport costs 8 denarii. You will pay the remaining 9 denarii . . . ought to be given (?) . . . nine pounds (?) . . . Farewell.'

---

[18] This was intended for the wax seals of the witnesses to a legal document, but the tablet was evidently not used for this purpose, or has been reused.

Notes:

1 There is a trace between the first and second **l** in **bello** which might or might not be a letter. The scratches on the wood show that this overlies an earlier text.

3 The correct reading is almost certainly **acceperunt**.

4 The word at the end of the line presents particular difficulty. Of the first three letters of **solues** only the **o** is certain. There is a clear high horizontal which has to be ignored if the first letter is read as **s**. The third letter might be **p**, and there is another apparent high horizontal which is discounted. The attraction of reading the word **solues** (from the verb **soluere** 'to pay') is obvious if the word '**denarios**' occurs twice in lines 4–5.

By using the images shown in Figure 1.3, we can see which of the scratches belonging to this text actually penetrated the wax to the wood so as to be visible under the different angles of lighting. We can also attempt to deduce which scratches did not penetrate or are not visible, and which are the scratches that belong not to this text but to an earlier text. The transcription below prints those letters and words which we think we might have been able to read without knowledge of the version on wax, but this is subject to some uncertainty in all lines except line 1 (where some letters appear more clearly on the wood than on the wax, no doubt because of the different quality of the photographs). The scratches which we have not been able to transcribe are confused and confusing; they probably belong to more than one text, and show significant differences when the tablet is illuminated from different angles:

albinus **bello suo salutem**
[traces only]
erunt in uec**turas**
ari**os oc**to
ari**os nouem qua**
am d**ebeb**

III. Word by Word Decipherment

Tablet 974

Our third example illustrates the difficulty of working from conventional photographs or unprocessed images of stilus tablets. We have a surface on which the inscribed text is confused by discoloration, wood-grain, and traces of at least one other text, which may or may not have been aligned with it (Figure 1.4). Other figures, reproduced in Chapter 2 (Figures 2.1, 2.2, 2.5, 2.6, 2.13, and 2.14), illustrate various aspects of the process: image capture of the surface of the tablet at 0° and 180°, with many other angles being possible in principle (Figure 2.13(a)); removal of the wood-grain by a simple process of Fourier transform (Figure 2.5); then the composite image, in which 0° and 180° are combined and the wood-grain is removed (Figure 2.5); next the highlighted section with stroke-detection using shadows and highlights (Figure 2.13(b)); finally the application

(a)

(b)

Figure 1.4 Vindolanda stilus tablet 974: (a) illumination from left; (b) illumination from right.

of phase congruency, which yields a variety of images (Figure 2.14). We must emphasise that they show different features of the text, and that in attempting to see and read the text, we actually negotiate our way round the different images and their constituent parts.

In thus negotiating our way, we feel somewhat like cells of the Earth, the computer in the *Hitch Hiker's Guide to the Galaxy* which was so complex that it required an organic matrix. This is provided by the two of us, and in what follows we will reconstruct the way in which we progressively read and understand an incised text like Vindolanda tablet 974. It has been drawn here (Figure 1.5) for convenience like the page of a modern book, black letters on white, in a reversal of the original, which was white on black, with the letters scratched into white wood through black wax. This wax has since disappeared, and we are left with the scratches, euphemistically called 'incised text'. How then do we 'read' them?

The great papyrologist Herbert Youtie read papyri scientifically; the adverb is deliberately chosen, since he formed hypotheses and tested them.[19] There is a story[20] about another master of palaeography, M. R. James, who catalogued all the western manuscripts in Cambridge, but is now better known for his *Ghost Stories of an Antiquary*. His own handwriting, appropriately enough, was illegible. His colleagues once received an

---

[19] See above, pp. 8–9.
[20] Told by George Lyttelton in Hart-Davis (1978), 14.

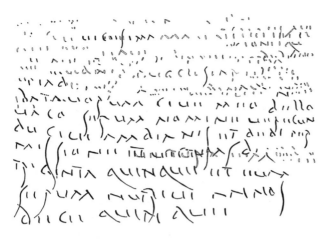

Figure 1.5 Tablet 974, drawn by R. S. O. Tomlin.

a	⋀	i	ι	q	⋋
b	⋌	k	⎮	r	⌐
c	⊂	l	⟍	s	⟋
d	⊲	m	⋀⋏ ⋯		
e	ιι	n	⋒		
f	⎰	o	◌	t	⊤
g	⊂̦	p	⌐	u/v	⊔
h				x	

Figure 1.6 Tablet 974: table of letter forms, drawn by R. S. O. Tomlin.

invitation to dinner: 'we *guessed* that the time was 8 and not 3, as it appeared to be, but all we could tell about the day was that it was not Wednesday'.

As it happens, this was good palaeographical method. Not a letter-by-letter transcription like sleep-walkers, but a wakeful testing of possibilities in the light of other knowledge. If you know the letter is an invitation to dinner—something given—you can exclude 3 o'clock thanks to your knowledge of the English dinner-time, and since you know there are only seven days in a week, the absence of the distinctive letter 'W' is a compelling argument from silence, like the letter 'S' is in our own stilus tablets, for it is a tall, powerful, sinuous letter which extends over two lines. Its corresponding drawback is that it often persists from an earlier text, and is thus unrelated to what we are trying to read.

Figure 1.5, the drawing of Vindolanda tablet 974, is incomplete since it represents work in progress. We have been forging a chain of interlocking hypotheses, but the last link has not been closed. Our starting point was the photograph[21] published by the excavators in their report, which gave us hope that the text could be read one day. In one of our copies there is evidence of this hope, the pencilled transcript of the last two lines. We might say, at this point, that one often starts from the end: the end of the whole text, the end of individual words. To illustrate this process, the lines in Figure 1.5 have been numbered from the *bottom* of the 'page'. Later, when we were given the original of the published photograph, we were able to read lines 1–3, the last three lines. Circumstances favoured us: these lines were not 'contaminated' by traces of an earlier text. This gave us most of the alphabet (Figure 1.6). In theory we might have deduced it from the text alone, like Sherlock Holmes reading *The Dancing Men* cipher first time off, but fortunately for us we had already seen other examples of Roman handwriting, and indeed other stilus tablets of the same date within the half-century AD 75–125.

As we have said, letter forms can be ambiguous at first sight.[22] In stilus-tablet texts 'e' and 'u', for example, can look the same since they are distinguished by the right-hand curve of the first down-stroke of 'u', which often disappears into the horizontal wood-grain. And although it was easy to read **nutriui** in line 2, there is a double down-stroke for 't' which has not been drawn, but is visible in the photograph. It may be the trace of a previous text, or the writer may have written 't', then erased it, and written it again; whatever the case, the clear reading of the rest of the letters in this word leaves little doubt that **nutriui** is indeed the correct interpretation of these signals. In these three lines we could already see that the spelling of two words had been influenced by the spoken language: the writer wrote **dece** for *decem* ('ten'), because the final 'm' was hardly sounded in speech, and he wrote **serum** for *seruum* ('slave') because 'u' and 'w' (literally 'double u') were indistinguishable in sound. Both these peculiarities armed us for the more difficult bits which lay ahead as we worked up the text, where we were aided by the improved images produced by our scientific colleagues. They reduced background 'noise', just as an observant drawing of an archaeological object is clearer than a photograph, and they distinguished deliberate stilus cuts from discoloration and casual damage.

At this point we could read the words '. . . thirty-five; and I have kept the slave for fifteen years'. We now had two hypotheses to test, as to what *type* of text it was: not a dinner invitation, clearly, but perhaps a deed of sale (vouching for the goods, one slave little used), or a deed of manumission (the freeing of the said slave). Working from the improved images, we could next read **missione**, in line 4. Naturally we then looked for *manu* in the line above, for 'manumission'; but there was too little space after the word **dedi** ('I have given'), and there was the recognisable descender of **r**, which is another helpful letter like 's', and also 'b' and 'q', because it occupies so much vertical space. So we could now read **et dedi permissione(m)**, 'and I have given permission', but then we were stuck. Moving forward in line 4, we could see that one text was written locally over another, indicating not a distinct text, but an erasure in the original, followed by a correction. The same letters were repeated, but they were 'out of sync' with each other. After quite a time, we decided the word must be **uecturas**, a word also found in Vindolanda tablet 836 (above), where it apparently means 'transport' or perhaps 'transportation charges'.

These references to 'permission' and to 'transport' made our text about a slave sound less of a legal document and more of a business letter. Perhaps the writer was sending a trusted slave on a journey. But what about 'thirty-five', and what has this got to do with 'transport'? We cannot find 'years' (*annorum*) at the end of line 4 — which might be the slave's age — nor can we find 'days' (*diebus*) — which might be the length of the journey, there and back. Unfortunately there are two texts here, one on top of the other, and they are both fragmentary. So we are still baffled by them.

But working back from **et dedi per** at the end of line 5, we could now recognise our good friend **serum** again in line 6, and still more helpfully, the first 'formula' we had met, the word **nomine**, 'by name'. This is like finding the letter 'q', which is bound to be followed by 'u'. After **nomine**, you can expect a personal name. Here again, the descender of **r** was helpful; there was **c** rising above the line, and the name began with either **e** or **u**, followed again by either **e** or **u**. The name Verecundus was the obvious candidate. Although it is a Roman name, it is what German classical scholars call a 'Deckname', a cover-name, since it conceals a Celtic name-element, and was therefore popular in Britain and Gaul. Two mistakes prevented us from recognising it more quickly than we did: at the beginning of line 5, we had read the preposition *de* (not -**du**), and after it *uico*, another formula: 'from the village of . . .'. But we now realised that 'de' must be -**du**, and that the writer, once again, had omitted the final '-m' of an accusative case-ending.

We now had a slave called Verecundus, who was apparently **ciui**, a 'citizen' of . . . somewhere. (We were assuming that **ciui** was an error for *ciue*, the accusative *cive*(*m*).) Thanks to the enhanced image, we could see that the place name began with a combination of **a** and **m** (**m** looks much the same as **aa** in this script), and that it ended with -**nis**. Mid-way, there was a **b** or **d** (another pair of letters difficult to distinguish when they are incomplete): **ambianis**, modern Amiens, satisfied us both, but we were uneasy that a slave should be 'citizen' of a town or tribe. We resigned ourselves to a search for parallels, still unfulfilled, and even wondered whether to read **cum** ('with the Ambiani') instead. But no; this seemed to be precluded by the traces visible.

At the beginning of line 6 was apparently the formulaic **uico** ('village'), at least if we assumed that one stroke was surplus to requirement, that it was a descender from line 7. Otherwise the reading would be **uaco**, which is not a Latin word at all. But fortunately it occurred to one of us that the last word in line 7 began with **be-**, and that the space and surviving traces (which have been over-restored in Figure 1.5 to illustrate what we *thought*) suited the reading **bello**. Before it there were two short words, fortunately short, since only the middle band of their incised letters survived: **ciue meo**, 'my fellow-citizen', was quite an easy reading, and this fitted **bellouaco**, which is the name of a Gallic tribe. Here the chain of hypotheses, as it were, closed for once. Neither of us

knew this at the time, but when we resorted to works of reference that evening, we found that the Bellovaci and the Ambiani were neighbouring north-Gallic tribes, both in the region of modern Amiens.

This was cheering: it was a real coincidence, not one which we had made for ourselves. That said, we must now admit that the previous word, the first in line 7, **batauorum**, is read by only one of us. The other does not accept it. Is collateral evidence any help here? The Ninth Cohort of Batavians was the battalion which garrisoned Vindolanda, and **batauorum** occurs often in the Vindolanda tablets. But neither of us, the optimist nor the pessimist, can read **cohortis** ('cohort'), which should accompany it, in line 8. Its absence is like that missing 'W' in the dinner invitation: Wednesday is out, and so is the battalion. Its absence casts doubt on **batauorum** as well.

This is almost as far as we have got, and it may be the end of the road. The further up the 'page', the worse it becomes. There are now *two* texts, if not three, and the band of lettering which survives has become narrower. It is also difficult to separate the lines of the different texts, whether by horizontal slope or vertical alignment, but these may be the only criteria. At the top of the 'page', optimistically emphasised in Figure 1.4, we think we can read **uicesima**, 'one-twentieth' or '5 per cent'. This might be another indication of the type of document, since one of its meanings is the tax levied on the manumission of slaves. Perhaps we will have to return to that hypothesis of manumission, after all.

We conclude with our working transcript and translation:

traces of 4–5 lines
Batauorum ciue meo Bello-
uaco ser(u)um nomine Verecun-
du(m) ciu(e) Ambianis et dedi per-
missione(m) et uecturas (*over* **uecturas**) . . .
triginta quinque et eum
ser(u)um nutriui annos
dece(m) quinque

'. . . of the Batavians(?) . . . my fellow-citizen of the Bello-vaci [name and verb lost] a slave called Verecundus, citizen(?) at Amiens. And I have given permission and travel-expenses(?) . . . thirty-five; and I have kept that slave fifteen years.'

It is incomplete, and only 'work in progress'. The emphasis should be on 'progress', since image enhancement has taken us further along an interesting road, and given us hope for the future. At least we are not alone. That story of the dinner invitation is capped by another about the handwriting of a legendary hostess, Lady Colefax: 'the only hope of deciphering *her* invitations, someone said, was to pin them on the wall and *run past them*'.[23]

---

[23] Rupert Hart-Davis in Hart-Davis (1978), 17.

This too is good scientific method, a primitive way of combining subtly different images seen in quick succession from different angles. But since our own mathematics go no further than O levels passed during the Mesolithic, we must leave the explanations to Brady and his colleagues (below). The story is only worth quoting because it catches something of the excitement and the mystery of phase congruency, at least for us, when the text 'moves' and letters jump out of the shadows.

References

BAGNALL, R. S. (1997), 'Imaging of papyri: a strategic view', in A. K. Bowman and M. Deegan (eds.), *Imaging Documents, Literary and Linguistic Computing* 12/3: 153–7.

BIRLEY, A. R. and BIRLEY, R. E. (1994), 'Four new writing-tablets from Vindolanda', *ZPE* 100: 431–46.

BIRLEY, E., BIRLEY, R. and BIRLEY, A. (1993), *The Early Wooden Forts: Reports on the Auxiliaries, the Writing Tablets, Inscriptions, Brands and Graffiti*, Vindolanda Research Reports 2 (Bardon Mill).

BOWMAN, A. K. and DEEGAN, M., eds. (1997), *Imaging Documents, Literary and Linguistic Computing* 12/3.

BOWMAN, A. K. and THOMAS, J. D. (1983), *Vindolanda: the Latin Writing-Tablets*, Britannia Monograph 4 (London).

BOWMAN, A. K. and THOMAS, J. D. (1994), *The Vindolanda Writing-Tablets (Tabulae Vindolandenses II)* (London).

BOWMAN, A. K. and THOMAS, J. D. (2003), *The Vindolanda Writing-Tablets (Tabulae Vindolandenses III)* (London).

BOWMAN, A. K., BRADY, J. M. and TOMLIN, R. S. O. (1997), 'Imaging incised documents', in A. K. Bowman and M. Deegan (eds.), *Imaging Documents, Literary and Linguistic Computing* 12/3: 169–76.

CROWTHER, C. W. (1997), 'Imaging inscriptions', in A. K. Bowman and M. Deegan (eds.), *Imaging Documents, Literary and Linguistic Computing* 12/3: 163–7.

CUNLIFFE, B. (1988), *The Temple of Sulis Minerva at Bath*, vol. 2, *The Finds from the Sacred Spring*, Oxford University Committee for Archaeology, Monograph 16 (Oxford).

FINK, R. O. (1971), *Roman Military Records on Papyrus* (Cleveland).

HART-DAVIS, R., ed. (1978), *The Lyttelton Hart-Davis Letters*, vol. 1 (London).

LALOU, E., ed. (1992), *Les tablettes à écrire de l'antiquité à l'époque moderne*, Bibliologia 12 (Turnhout).

MARICHAL, R. (1992), 'Les tablettes à écrire dans le monde romain', in E. Lalou (ed.), *Les tablettes à écrire de l'antiquité à l'époque moderne*, Bibliologia 12 (Turnhout): 165–85.

OBBINK, D. (1997), 'Imaging the carbonized papyri from Herculaneum', in A. K. Bowman and M. Deegan (eds.), *Imaging Documents, Literary and Linguistic Computing* 12/3: 159–61.

SPEIDEL, M. A. (1996), *Die römische Schreibtafeln von Vindonissa*, Veröffentlichungen der Gesellschaft pro Vindonissa XXII (Brugg).

TERRAS, M. (2000), 'Towards a reading of the Vindolanda stylus tablets: engineering science and the papyrologist', *Human IT* 2/3: 255–71.

TOMLIN, R. S. O. (1988), 'The curse tablets', in B. Cunliffe (ed.), *The Temple of Sulis Minerva at Bath*, vol. 2, *The Finds from the Sacred Spring* (Oxford): 59–277.

TOMLIN, R. S. O. (1994), 'Vinisius to Nigra: evidence from Oxford of Christianity in Roman Britain', *ZPE* 100: 93–108.

TOMLIN, R. S. O. (1998), 'Roman manuscripts from Carlisle: the ink-written tablets', *Britannia* 29: 31–84.

YOUTIE, H. C. (1973), *Scriptiunculae*, vol. 1 (Amsterdam).

CHAPTER 2

Shadow Stereo, Image Filtering, and Constraint Propagation

Michael Brady, Xiao-Bo Pan, Veit Schenk, Melissa Terras, Paul Robertson &
Nicholas Molton

Introduction

The discovery and decipherment of Latin writing-tablets, including texts incised on wood and lead, from Vindolanda (near Hadrian's Wall), Carlisle, Bath, and London, during the past three decades has fundamentally changed the historian's picture of Roman Britain and provided by far the most important source of new evidence for the development of the Latin language. The texts deciphered and published since 1974 form a fundamental research archive of a kind which, for all practical purposes, was quite unparalleled before the discoveries mentioned above. The work described in this article aims to develop a system to help the historian read such stilus tablets by improving their legibility. There are hundreds of existing tablets, while archaeology continues to increase the stock of material to be read.

Stilus tablets may be one of the most important documentary sources, but they are also the most difficult for expert Roman historians to decipher, and this forms the basis of our collaboration, and of the computer vision challenge. Figure 2.1 shows a stilus tablet which has been imaged by a specialised CCD and camera body positioned vertically about 75 cm above the tablet, which generates high-quality images up to 7000 pixels square. Note that the variations in brightness of illumination across the surface of the stilus tablet are deliberate, as we describe below.

It is clear from Figure 2.1 that, as well as being huge, the rough surface of the stilus tablet has undulations between the lines corresponding to wood-grains (typically with an amplitude of 3 mm). As a result, the image has poor signal to noise and it is heavily textured. The signals that are of interest — the incisions — are of low brightness contrast. The dense wood-grain lines of the silver fir, of which stilus tablets were mostly made, make reading even more difficult by breaking up some of the incised strokes and by further complicating the surface reflectance of the tablet. Finally, stilus tablets are often quite badly stained and pitted. For all of these reasons, stilus tablets pose immense difficulties for conventional 2D image analysis.

However, the key attribute of the writing is that it is *incised*, suggesting that three-dimensional image analysis techniques be used, a point that we address in the next section. If the surface of the stilus tablet can be determined, as well as incisions into that surface, then, for example, since discolorations lie only on the surface of the tablet they are of no interest and can be ignored.

(a)

(b)

Figure 2.1 (a) Vindolanda stilus tablet 974, measuring about 12 × 8 cm; the writing surface is recessed to a depth of approximately 3 mm. (b) Magnified portion of the window indicated in (a), containing the letter M.

On the other hand, any 2D analysis is likely to accord particular attention to intensity changes, and the intensity change across a stain is often far higher than that across an incision.

The incisions of interest typically measure 0.5 mm across and are of varying but shallow depth, to a maximum of 1 mm but tapering to zero. It follows that any attempt to separate incisions on the basis of depth differences faces a severe practical difficulty: an incision that is 1 mm in depth and is viewed from above from a height of 75 cm generates a relative depth of the 'signal' with a maximum value of 1:750, while the surface roughness 'noise' can be as much as 20 per cent of this figure, and surface undulations have an amplitude three times as great as the incision depth. It follows that absolute or relative range estimation confronts a difficult challenge. Finally, the specialised CCD camera that we use imposes some constraint on the viewing geometry: it is difficult to vary the viewing geometry from above the sample, but there is scope for control over the lighting. It is this scope that we exploit.

Three-Dimensional Analysis of Stilus Tablets

Several methods have been developed in 3D computer vision. Which, if any, is the most suitable for the task in hand? For the purposes of this article, it is useful to distinguish between techniques that estimate range (the distance from the system to a point in the scene) *directly*, and those that *infer* depth from one or more images. The latter, including binocular stereovision, are related to the capabilities of the human visual system. Practical considerations are that the 3D image analysis technique chosen should not only be capable of detecting incisions, but also be portable (since it has to be taken to the sites — for example museums — where the stilus tablets are to be found) and relatively inexpensive, since research budgets for this kind of work are modest.

Direct Ranging Techniques

The article by Wallace (Chapter 9) is an excellent overview of direct ranging techniques. One class of direct ranging technique uses 'time of flight': electromagnetic energy of a suitable frequency is directed at the scene; the times of departure and arrival of the reflected energy are measured accurately, and from their difference, as well as knowledge of the speed of propagation of the energy, the range to the target range can be calculated. The beam is then scanned vertically and horizontally to form an image. Several such devices have been developed, even commercialised, and more advanced devices are under development (see Chapter 9). Unfortunately, since light moves so quickly (approximately 230 metres in a microsecond), even if a system can sample at nanosecond rates, the distance that the light moves is 230 times the depth of an incision. There are ways to reduce this problem, for example by using a high-performance circuit to measure the phase delay of a returned wave relative to that being emitted, phase delay being linearly related to range. Even the impressive experimental systems described by Wallace struggle to get near to the required resolution, and so we need to look elsewhere.

A second approach might be to use confocal scanning optical microscopy (CSOM), as described in Wilson (1990). CSOM is attractive in the present context since it is a direct range sensing method which is appropriate for measuring fine incisions on a surface. Unfortunately, CSOM devices are currently quite expensive, bulky, and slow. A small practical point: even if CSOM devices were inexpensive and portable, curators would need to be convinced by safety aspects of illuminating priceless artefacts with laser light. Nevertheless, the confocal technique is developing rapidly, so its use may soon become a feasible proposition, in which case we would gladly use it. However, we deemed it unsuitable for our project.

Finally, there are a variety of techniques referred to collectively as 'structured light' in which a known illumination pattern (e.g. closely separated parallel stripes) is projected onto a scene, which is viewed from an offset angle. The illumination pattern is chosen in such a way that it is easily seen in the image. Accurate calibration of the camera/projector geometry enables the appearance of the illumination (e.g. displacement of a stripe) to be converted to a depth. Structured light techniques can yield accurate range data. However, to date they have worked best for estimating the gross geometry of a smooth shape, not for estimating the depth of fine incisions on a rough surface. Often, structured light systems yield only sparse information (e.g. no information is explicit in the spaces between parallel light stripes), but this is not acceptable in the current project, where the depth variations to be detected are shallow and, since the writing is difficult to interpret even on ink tablets, all the information contributing to a perceived incision should be extracted. Again, we rejected this approach for our project.

Inferred Range

Binocular stereo (also known as *stereo vision*[1]) exploits the differences between images received at two nearby imaging devices (e.g. the eyes) to infer the 'disparity' of points in a scene. Stereo disparity is inversely related to depth, the precise relationship necessitating accurate calibration of a system. Over the past twenty years, a comprehensive theory of binocular stereo has been developed, and some efficient algorithms have been invented to automate calibration and to compute disparity. However, even though binocular stereo has now developed to the point where disparity maps can be computed in real time on moderately powerful computer hardware, it is hard to apply the technique to our project. This is because, as we noted above, the change in disparity (depth) when descending through an incision is very small relative to the noise variation in disparity of the tablet surface. Use of binocular stereo would necessitate an extremely precisely engineered and calibrated system, increasing cost and reducing portability. In like manner, there has been considerable progress over the past thirty years in developing the computation of 3D structure from monocular or binocular motion — however, the same limitations apply as for binocular stereo.

The final candidate technique from the literature of computer vision is known as *shape from shading*. It is well known that variations in brightness generate a strong impression of three-dimensional structure. Horn, in particular, has analysed the process of image formation: if the reflectance function of a surface is known, as is the relative placement of the viewer and the 'sun' illuminating the scene, then the 3D geometry of the scene can be estimated from the brightness variations in a single image.[2] Several impressive algorithms have been developed to compute shape from shading. In many practical applications, including the current one, precise knowledge of the surface reflectance is not available. A particularly interesting development of shape from shading is *photometric stereo*.[3] A fixed camera takes images of the same scene, which is illuminated

---

[1] Faugeras (1993).

[2] Horn (1972, 1986).

[3] Woodham (1980, 1994).

in succession by one of three or more lights (whose directions are known), yielding a set of geometrically aligned images of the scene. In principle, from two such images it is possible to infer surface orientations, but this still requires accurate knowledge of surface reflectance. Using three or more images it is possible to do without knowledge of reflectance, but careful calibration is needed.

Shape from shading is the one technique that conforms to the 'fixed viewing geometry, some scope for varying illumination' constraint, but its principal drawback is that systems constructed to date have been designed to estimate gross surface geometry, and do not facilitate finding the incisions. Second, and more pertinent to the current project, it is known that while shape from shading works well for convex portions of shapes, concavities pose severe challenges. In particular, the simple light reflection model that is central to shape from shading no longer applies, since there are numerous internal reflections in the interior of a concavity.[4] We (reluctantly) concluded that a novel approach was required.

(a)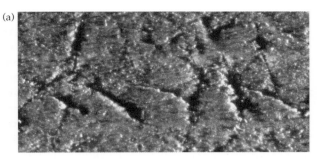

(b)

Figure 2.2 The two images are of the same fragment of a stilus tablet: (a) is taken from a relatively low elevation (25° above the plane of the stilus tablet), whereas (b) is at a relatively high elevation (75°). Note, for example, for the indicated incision, the substantial cast shadow and thin highlight in the upper case, and the converse in the lower case.

Shadow Stereo

Introduction

Our new approach came from watching Roger Tomlin show us how he examines a stilus tablet: he balanced it on his upturned palm, at eye level, and then slowly rotated the tablet about a horizontal axis orthogonal to the line of sight. We interpret this as using low raking angle light, close to the plane of the stilus tablet, and changing the elevation. Furthermore:

1 under low angle illumination, the edge of an incised stroke nearest the light causes a cast shadow, next to a highlight which is generated by the trailing edge of the incision;
2 the width of a cast shadow increases as the elevation of the lighting direction is lowered towards the plane of the stilus tablet, and the change in shadow extent is predictable as the elevation changes (for fixed azimuthal direction of illumination);
3 the width of the cast shadow can be regarded as a detectable signal in the image that *amplifies* the small incision depth;
4 surface coloration patches lying in the surface of the stilus tablet do not change in the way that shadows do; and
5 the shadows of interest are adjacent to a highlight.

These effects can be seen clearly in Figure 2.2. We are led to the following algorithm for detecting incisions.

Shadow Stereo Algorithm

Pick an azimuth light direction, then take a set of images at different elevations:

1 for each:

   a remove the wood-grain
   b apply a feature detection algorithm to locate shadows adjacent to highlights
   c use these to detect candidate strokes

2 detect those candidate strokes that move as they 'should' do

---

In the remainder of this section, we describe the image formation process and show how we choose the set of elevations for each azimuth direction. The hardest part of realising the shadow stereo algorithm is the feature detection required in step 1b. This is the subject of the next section.

Imaging Details

To achieve a reasonable resolution on the incisions that correspond to the strokes on individual characters, we use a medium format, high-resolution camera capable of generating images that are 7000 pixels square (we used a Fuji GX680 camera body with a Fujinon 1345/5.6 lens, and a Phaseone Powerphase 14 bit scanning back). The captured colour image is converted to 8 bit grey levels, so that each image is up to 49 megabytes in size. The design of the camera means that it is best held in a fixed position above the tablet so as to produce a view which

---

[4] Forsyth and Zisserman (1990, 1991).

always looks directly down onto the writing surface. The light source elevation is altered to cast shadows from the incisions of varying extents. Since the positions of the tablet and the camera remain fixed, there is no need to geometrically align the images from different light elevations. Illumination is provided by a set of fixed compact light sources. We find that fixing lights a distance of 75 cm from the leading edge of the tablet gives good results. Multiple individually switchable lights enable light to be provided from different elevations at the same azimuth. Different azimuths are produced by moving the tablet and camera together on a turntable.

The written part of the stilus tablet measures approximately 12 × 8 cm, so that each pixel corresponds to a tablet sample approximately 16.7–20 microns square. At this finest resolution, a typical letter occupies 290 × 250 pixels, while an individual stroke is 6–20 pixels across. Creating an image at this resolution gives a magnified view of the tablet, which, together with a rapid zoom facility, greatly improves legibility of the text.

Choosing Elevations

This section summarises the analysis presented in Molton et al. (2002). Correctly positioned lighting provides clear shadows, reduces the effect of noise, and generates the maximum amount of information about the incision structure. The light elevation needs to accentuate incisions and hide surface noise. Such noise is common, appearing mostly as shallow hollows produced as a result of the texture of the wood, shrinkage, minor surface damage, and surface discoloration. Since the shadow stereo algorithm aims to differentiate between incisions and surface discolorations using shadow motion, the induced motion should be as large as possible (see Figure 2.3).

When a straight incision is viewed under light at an elevation ε and relative azimuth α_i, the extent s of the shadow is given by:

$$s = \frac{2w_i \cos \alpha_i \tan \theta_i}{\tan \varepsilon + \cos \alpha_i \tan \theta_i}$$

where w_i is the incision width and θ_i is the back face angle (see Figure 2.3). Differentiating this equation with respect to light elevation ε gives:

$$\frac{\partial s}{\partial \varepsilon} = \frac{-2w \tan \theta_i \cos \alpha_i}{\cos^2 \varepsilon (\tan \varepsilon + \cos \alpha_i \tan \theta_i)}$$

This equation suggests that shadow extent varies more when small changes are made to a low (raking) light elevation, indicating that a lower light elevation should result in an increased chance of incision detection. This explains mathematically what Roger Tomlin already knew perfectly well! Of course, if the light elevation is too low, it may result in the shadow boundary appearing very close to the top of the incision, where the chance of tablet damage is high. A shadow falling here will provide little useful information about the shape of the incision. Response to shallow indented noise is also high for very low raking light. We calculated the theoretical response to different elevations, for different incision models. These experiments are described by Molton.[5] As a result, for each azimuth direction around the stilus tablet, we use two elevations at approximately 5° and 40°.

Correlation

Step 2 of the shadow stereo algorithm extracts just those candidate strokes that move as they 'should'. From the above analysis, we can predict the change in shadow extent given that we know the two elevations used in the algorithm. Let this be denoted by μ. Then, denoting the elevations as 'high' and 'low', we would expect a high value for the expression:

$$SIM\left(I^{low}(\mathbf{x}), I^{high}(\mathbf{x} + \mu)\right)$$

where SIM is a suitable similarity measure, such as correlation, mutual information, or correlation ratio. Figure 2.4 shows visually the value of this expression using the correlation of phase congruency maps (as detailed in the next section).

Feature Detection

Removing the Wood-Grain
The wood-grain often has relatively high contrast, and is distracting both to the historian examining the tablet and to algorithms attempting to enhance the image. Stilus tablets were recessed so that the wood-grain always runs close to the horizontal direction. We developed two algorithms to remove the wood-grain. The first used a conventional band-pass filter to remove a range of frequencies (corresponding to the wood-grain) in the vertical direction. This conventional algorithm works quite well. However, it also tends to remove some of the important horizontal incisions in the image. This makes it harder to discriminate between similar letters with horizontal strokes, for example 't' or 'p'. The filter's cut-off frequency must be set up manually, according to the image resolution, and the approximate period of the wood-grain.

The second method exploits a number of photometrical observation properties of stilus tablets in order to remove the intensity variation in an image attributable to

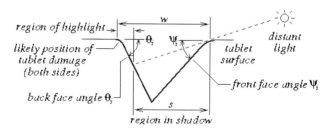

region of highlight →
likely position of tablet damage (both sides)
back face angle θ_i
θ_i
Ψ_i
w
tablet surface
distant light
front face angle Ψ_i
s
region in shadow

Figure 2.3 Geometry of shadow formation.

5 Molton et al. (2002).

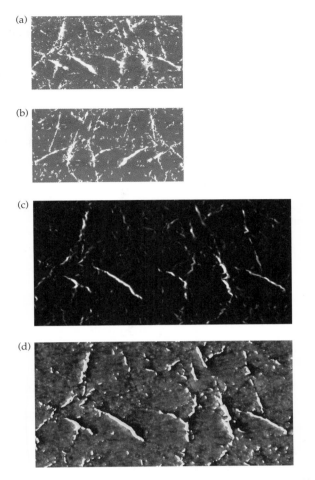

Figure 2.4 The thresholded correlation of phase congruency (PC) feature values for the two images shown in Figure 2.2. (a) The PC map for Figure 2.2(a). (b) The PC map for Figure 2.2(b). (c) The correlation as defined in the text. (d) The correlation overlaid on the image.

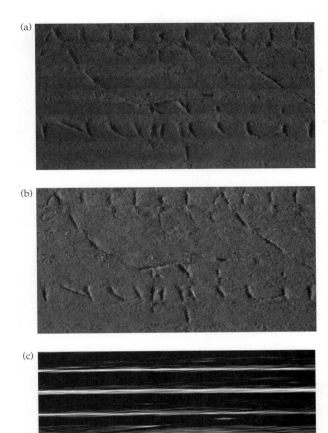

Figure 2.5 Wood-grain removal. (a) Original image fragment. (b) Result of removing wood-grain using the reflectance algorithm described in the text. (c) The wood-grain detected by the method.

wood-grain. It incorporates the following assumptions (conventionally, the x-axis runs horizontally):

- wood-grain is predominantly approximately parallel to the x direction
- the surface normal $\hat{n}(x, y)$ varies relatively quickly in the y direction because of the wood-grain, but more slowly in the x direction, so that $\hat{n}(x, y) \approx \hat{n}(y)$
- the reflectance of the stilus tablet surface is approximately Lambertian
- the local colour (albedo), essentially the colour of the wood, remains constant.

Let $I(x, y)$ be the intensity of the image at point (x, y), and denote by θ_y the angle between the light direction and the tablet surface normal. We find:

$$\sum_x I(x, y) = c \cos \theta_y$$

where c is a constant. By setting θ_y to 0, that is, by adjusting the intensities of image rows so that they are all the same, the wood-grain can be removed. In practice, the large images were divided into vertical segments. The very dark and very bright points, which correspond to the shadow and highlights in the image, are identified in the histogram and left out of the above calculation, to ensure that the third and fourth assumptions are met. An important advantage of this method is that it does not require any parameters to be set. Figure 2.5 shows an example of the algorithm's performance.

Finding Shadows Adjacent to Highlights

Though there has been a great deal of work in image analysis on feature detection, the vast majority both of mathematical analyses (example: anisotropic diffusion) and algorithms implicitly assume that a 'feature' corresponds to a step change in signal intensity. Such algorithms perform poorly when confronted with other kinds of signal change. Figure 2.6(a) shows a typical illuminated incision on a stilus tablet, and Figure 2.6(b) shows the intensities along the line AB indicated in Figure 2.6(a). The shadow region is clearly visible as the intensity valley between A and B, and the highlight is equally visible as the adjacent intensity peak. Critically, while the important transition from shadow to highlight is evident to the human eye, it is far from being a step change.

We found that the feature detectors used most frequently in image analysis gave very poor results

(a)

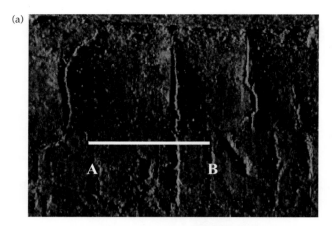

(b)

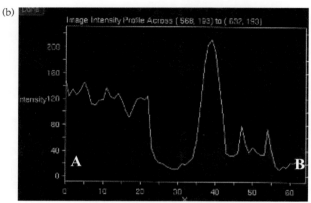

Figure 2.6 (a) A fragment of a stilus tablet. The intensities along the line indicated AB are shown in (b). Note the deep valley corresponding to the shadow, adjacent to the peak, corresponding to the highlight.

on the stilus tablets. Fortunately, there is another approach. Morrone and Owens (1987) observed that all commonly occurring one-dimensional signal changes — steps, ramps, thin bars, dots — correspond to signal locations at which the local Fourier components are all in phase. Furthermore, they found that even complex intensity changes often correspond to points of maximum phase 'congruence'. They established the link between a signal feature to the point of maximum phase congruency. They also proved that their definition of phase congruency is a normalised measure of a well-known computation, the *local energy*. This provides an important explanation of the concept of the local energy: peaks in the energy function correspond to feature points where phase congruency is a maximum. Being a measure of phase 'congruence', PC (or local energy) has the advantage of not being sensitive to the image contrast and brightness.

Local Energy and Phase Congruency (PC)
Let $f(x)$ be a one-dimensional signal. It can be reconstructed from its Fourier spectrum by

$$f(x) = \int_{-\infty}^{\infty} a_\omega \cos(\omega x + \phi_\omega) d\omega$$

where a_ω is the amplitude and $\omega x + \phi_\omega$ is the phase offset. Since steps, roofs, and so on all correspond to points in

a signal where the components of the spectrum are in phase, the phase congruency $PC(x)$ at each point x in the signal is defined as:

$$PC(x) = \max_{\theta \in [0, 2\pi)} \frac{\int_{-\infty}^{\infty} a_\omega \cos(\omega x + \phi_\omega - \theta) d\omega}{\int_{-\infty}^{\infty} a_\omega d\omega}$$

The θ that maximises this expression for the PC result represents the amplitude-weighted mean phase angle. By definition, PC is a dimensionless value between (0,1).

It is very awkward to compute PC directly from its definition. It is normally obtained from the local energy, computed as below, using the relationship:

$$PC(x) = \frac{E(x)}{\int_{-\infty}^{\infty} a_\omega d\omega}$$

The local energy can be obtained from the analytic wavelet transform, which is equivalent to convolving the signal with a pair of quadrature filters. An analytic wavelet ψ is a function whose Fourier transform is zero for non-positive frequencies:

$$\tilde{\psi}(\omega) = 0 \text{ if } \omega \leq 0 \text{ (} \tilde{} \text{ denotes the Fourier transform)}$$

Let f be the signal, then the result W of transforming the signal with ψ is the inner product

$$W = \langle f, \psi \rangle$$

In general, W is complex. The local energy E is the amplitude of the transform:

$$E = |W| = |\langle f, \psi \rangle|$$

Similarly, $p = Arg(W)$ gives the phase angle at which the phase congruency occurs, and will be used later to specify the feature type. The analytic function has zero phase. It does not change the signal's phase. Instead, it separates the signal's amplitude and phase. This is so for any analytic wavelet, so the question arises which one to use.

Gabor functions are widely used and are considered approximately analytic, if the non-positive frequencies are small enough. This requirement restricts the use of the Gabor filter: one cannot construct Gabor functions of arbitrarily wide bandwidth and still maintain a reasonably small non-positive frequency component.

The log-Gabor function is found to have better properties. It is analytic by definition. It has a long high frequency tail, which is useful for detecting fine features. The log-Gabor function is defined in the frequency domain as

$$G(\omega) = e^{\frac{-(\log(\omega/\omega_0))^2}{2(\log(\kappa/\omega_0))^2}} \text{ if } \omega > 0, \text{ and all zero otherwise}$$

where ω_0 is the filter's centre frequency. The term κ/ω_0 is held constant. A κ/ω_0 value of 0.74 will result in a filter bandwidth of approximately one octave, 0.55 will result in two octaves, and 0.41 will produce three octaves. Figure 2.7 is a plot of log-Gabor functions of

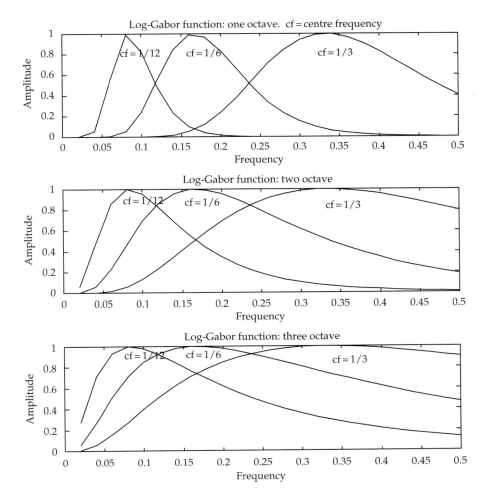

Figure 2.7 Log-Gabor functions of different bandwidths and centre frequencies.

different bandwidths and centre frequencies. The function's centre frequency and bandwidth are to be chosen according to the application.

Figure 2.8 shows the phase congruency of a sample signal calculated using log-Gabor functions and the Hilbert transform. The solid line of PC obtained from the log-Gabor function has peaks for every feature point, which are marked with *. Note, however, that the dotted line of the Hilbert transform has only two peaks, instead of four, in the second half of the signal. The reason is that the Hilbert transform of the 1D signal is a special case of the analytic wavelet transform where the whole positive frequency spectrum is evenly covered, unlike the band-pass filter as the log-Gabor functions. Therefore its time–space coverage is too narrow to produce all the feature peaks. We next discuss a method to recover all the feature points using phase angles.

Detecting Feature Type from Phase Angle
Consider the third sub-plot of Figure 2.8, which is the phase angle $p(x)$ obtained from the Hilbert transform of the sample signal. Each feature point corresponds to a specific phase angle. Figure 2.9 shows the feature type and the phase angle.

Combining the phase information with the PC (or local energy), it is possible to distinguish the peaks in the local energy (or phase congruency) function. More precisely, PC and phase can be combined in the following

way, which gives a new measure of the features, denoted as PC_f,

$$PC_f(x, \phi_i) = PC(x) \sum_i \lfloor \cos(p(x) - \phi_i) \rfloor^2$$

where $p(x)$ is the phase of the signal as defined in the previous section, and ϕ_i is the phase at the following feature types

$$\phi_i = \begin{cases} 0 & \text{peak} \\ \pi/2 & \text{up step} \\ \pi & \text{valley} \\ -\pi/2 & \text{down step} \end{cases}$$

i is one of the feature types listed above, and $\lfloor \rfloor$ denotes that the enclosed quantity is equal to itself when its value is positive and zero otherwise. PC_f can be used to detect any of the feature types, or, more importantly, any combination of them.

The usefulness of the above combination is illustrated in Figure 2.10, where different features were detected from the sample signal of Figure 2.8. The two feature points in the second half of the signal are clearly identified in the up-step feature. Even more features show up: for example, the peak of the first half of the signal and the long down slope in the second half of the signal.

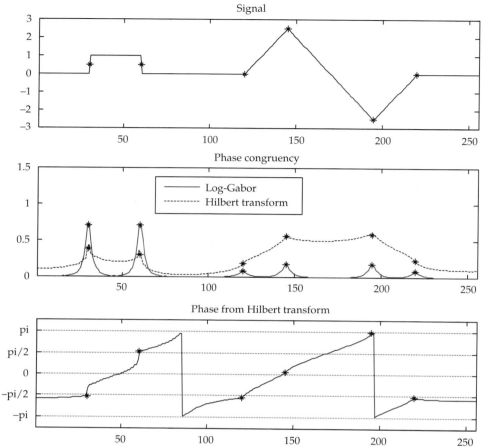

Figure 2.8 Phase congruency and phase of a sample signal.

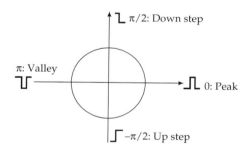

Figure 2.9 Phase angles of different feature type.

Extending PC to 2D

The local energy and phase congruency computation can be extended to 2D or 3D data by following Granlund's method[6] if it is assumed that the signals of interest have simple neighbourhoods, that is, vary locally only in one direction. With this kind of image data, it is possible to interpolate the local energy and estimate the orientation from a minimum of three energy outputs obtained from three symmetrical distributed directions: 30°, 90°, and 150°. The energy was computed in each direction with an analytic wavelet function constructed at this direction and extended with a spread function $\cos^2(\phi)$. In practice, six directions (0°, 30°, 60°, 90°, 120°, and 150°) were used to accommodate the complexity of the images.

For illustration, this method of extending PC to two dimensions is applied to the idealised image shown in Figure 2.11(a), with the result shown in Figure 2.11(b). The method also gives the approximate local signal orientation, as shown in Figure 2.11(c).

Multiscale Analysis

Phase congruency can be applied to an image at multiple scales and at different filter bandwidth. Figure 2.12 shows the scalogram of applying PC on a section of an incised tablet image. The first and second parts of Figure 2.12 show the tablet image and a signal from the image. The next three scalograms were obtained with three log-Gabor filters of different bandwidths. The horizontal axes of the scalograms correspond directly with the signal's horizontal axes. The vertical axes of the scalograms are scales ranging from 1 to 20, and correspond to the filter wavelength from 3 to 20. The scaling factor between the filters is 1.1. The scalogram shows the lifeline of a feature across scale. We can choose the most suitable scale and bandwidth to do a single scale analysis of PC or local energy, or multiscale analysis using Kovesi's method.[7] However, we have found that the performance of Kovesi's method depends on noise estimation. In the case of tablet images, we find that single scale analysis achieves better results.

---

[6] Granlund et al. (1994).

[7] Kovesi (1999).

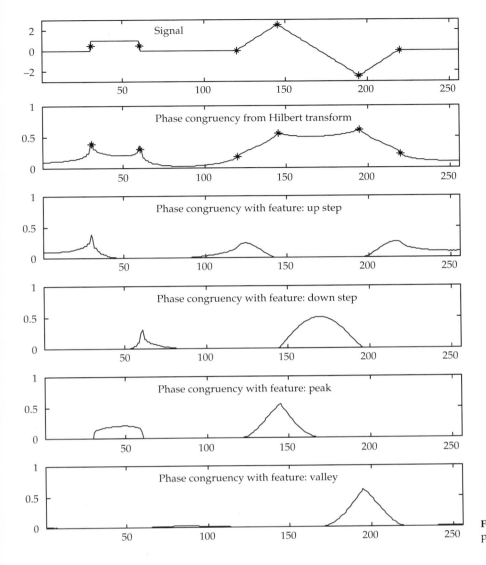

Figure 2.10 Detecting features using the phase angle.

Application to Stilus Tablets

The PC method described above has been applied to stilus tablet images in order to detect incisions. This was followed by hysteresis thresholding to binarise the result, so that the incisions are identified. An adaptive way of determining the high and low thresholds was used. The number of identified points of incision against various high and low thresholds was plotted for each processing region. The slowest changing and suitable points were used as the thresholds. Figure 2.13(a) shows an image illuminated from the right. The highlight, shadow of the incision, and the transition between them can be treated as peak, valley, and down-step features, and are detected and illustrated in different colours in Figure 2.13(b).

Filling in the Gaps

Although the results of the feature detection stage are encouraging (Figure 2.13a), it is immediately apparent that there are gaps in the detected strokes. Such gaps often, though not invariably, coincide with the wood-grain representation that was detected in Figure 2.5(c). This section outlines an algorithm for filling in such gaps, an algorithm that was originally devised to model

human perception of subjective contours.[8] The result of applying the algorithm to Figure 2.14(a) is shown in Figure 2.14(b).

The algorithm requires that stroke portions be identified, in particular their end points and the stroke portion direction. To do this, we applied a shape primitive identification algorithm which uses an adaptive region of interest. Each end point of each stroke portion detected in this way is assigned a likelihood that the stroke portion should be joined to one that is adjacent. If the end point is in a region identified as being wood-grain, the likelihood is increased. Proximity of two likely stroke portion end points, at which the stroke directions are roughly collinear, further increases the likelihood that the stroke portions should be joined. More precisely, a search is conducted in the local area of each end point for candidate partners to participate in a join. If a pair of ends has been established from both of them, then they are treated as a candidate pair. Each pair has a joint likelihood. An adjustable threshold was used on the likelihood to finalise the identification.

[8] Ullman (1976).

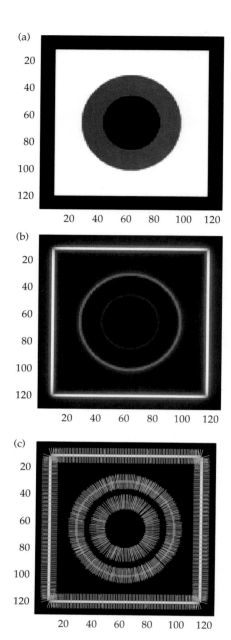

(a)

(b)

(c)

Figure 2.11 (a) Idealised test image. (b) Phase congruency. (c) Orientation vectors.

for $x(t)$ and $y(t)$ are defined for $t \in [0, 1]$ by

$$x(t) = A_x t^3 + B_x t^2 + C_x t + D_x$$

$$y(t) = A_y t^3 + B_y t^2 + C_y t + D_y$$

where $P1 = [x(0), y(0)]^t$, $P2 = [x(1), y(1)]^t$, $v1 = [x'(0), y'(0)]^t$, and $v2 = [x'(1), y'(1)]^t$. It is straightforward to show that

$$A_x = 2p_{1x} - 2p_{2x} + v_{1x} + v_{2x}$$
$$B_x = 3p_{2x} - 3p_{1x} - 2v_{1x} - v_{2x}$$
$$C_x = v_{1x}$$
$$D_x = 2p_{1x}$$
$$A_y = 2p_{1x} - 2p_{2x} + v_{1x} + v_{2x}$$
$$B_y = 3p_{2x} - 3p_{1x} - 2v_{1x} - v_{2x}$$
$$C_y = v_{1x}$$
$$D_y = 2p_{1x}$$

The magnitudes of $v1$ and $v2$ are two free parameters and are defined as

$$|v1| = |v2| = 2D/[1 + (\cos \varphi_1 + \cos \varphi_2)/2]$$

From Signal to Symbol

Introduction

The stilus and ink tablets are Latin texts, written by a small number of scribes in a population that was predominantly illiterate. The act of reading such texts, indeed the act of reading *any* text (for example this one) is a complex process of *interpretation* in which knowledge of the written form of the language, as well as the subject matter of the text, is mobilised in a way that is currently only partially understood. Reading is difficult to understand for the same reason that understanding any other skilful process is: we unconsciously mobilise knowledge of many different sorts. At the outset of the project, historians were aware of this, largely through the papers of Youtie, who was the first to attempt to describe the processes that papyrologists go through when transcribing and translating a text.[9] Equally, the perception of reading has been a major topic of research in perceptual psychology, particularly during the past three decades.[10] During the 1980s, McClelland and Rumelhart (1986) published an influential theory of reading in terms of parallel distributed processing. In the course of this project we have:

1 developed a knowledge-elicitation process that attempts to uncover, in terms of the Rumelhart–McClelland model of reading, what 'level' of knowledge is being mobilised by two highly skilled historians as they interpret a stilus tablet. Results show that the knowledge deployed is very different when the historians interpret a tablet *de novo* and when they are asked, some months later, to interpret the same tablet (at which point they have a strong idea what the subject matter of the tablet is);

Two methods were then investigated to fill the gaps between the broken strokes. The first is a variant to the algorithm of Rutkowski (1979) that models the stroke join using two circular arcs that are tangent to the gap ends and to each other, and, among such pairs, have minimum total curvature. The second represents the gap by a cubic polynomial to represent the missing segment, given the two end points and the directions in which the stroke is going at those points. The latter was used because of its simple implementation. We note that in the small angle case (that is, approximate collinearity), the cubic minimises the integral squared curvature of the gap-filling curve (see Figure 2.15). We denote the end points to be filled by P1 and P2, and the direction vectors $v1$ and $v2$. Let $P1 = (p_{1x}, p_{1y})^t$, $P2 = (p_{2x}, p_{2y})^t$, $v1 = (v_{1x}, v_{1y})^t$, $v2 = (v_{2x}, v_{2y})^t$. Then the parametric cubic polynomials

---

9 See above, pp. 8–9.
10 Gibson and Levin (1975).

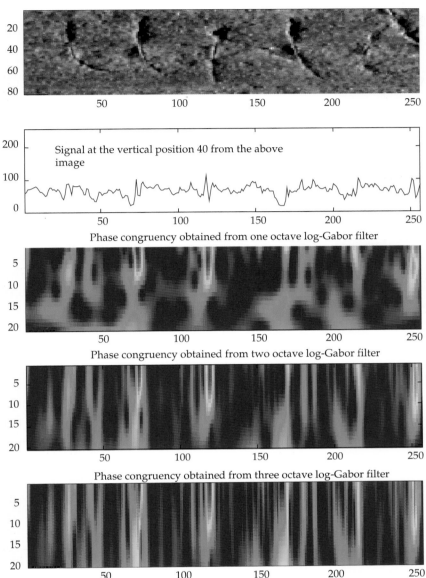

Signal at the vertical position 40 from the above image

Phase congruency obtained from one octave log-Gabor filter

Phase congruency obtained from two octave log-Gabor filter

Phase congruency obtained from three octave log-Gabor filter

Figure 2.12 Phase congruency at different scales.

2 developed a wholly new and unique resource for palaeographical research in the form of a corpus of words and letter shapes from a sample of Vindolanda material. This is of great significance for palaeography since a major underlying assumption of the field is that at any period there was a narrow range of writing conventions, reflecting the relatively small number of writers. Early indications using a computer-based analysis of the corpus challenge this underlying assumption. The corpus is currently being made available to scholars worldwide using the Internet;

3 adapted an alternative computer architecture, initially developed for interpreting aerial images (as foreshadowed in the project proposal), to help historians read writing tablets, based on: stroke data culled from the writing tablets, together with a representation of the written forms of Latin characters, and word and bigram frequency counts. Though the architecture has been designed to interact with historians in order to help them read tablets, including keeping track of alternative readings of a portion of the tablet, it has successfully read whole lines of stilus tablets entirely automatically.

Knowledge Elicitation

To aid in the complex cognitive and perceptual processes involved in the reading of such damaged, deteriorated texts, an appropriate knowledge-based computer program has been developed, firstly to try to make explicit the reading techniques that papyrologists use while deciphering such texts (which have so far remained implicit), and secondly to provide a tool to aid the papyrologists in recording the recursive hypotheses they develop in the transcription of the texts.

Papyrology is in essence a 'self consuming labor which leaves little or no trace of itself',[11] and the expertise of papyrologists, as with the expertise of any professional, is a valuable but surprisingly elusive resource.

[11] Youtie (1963), 11.

(a)

(b)

Figure 2.13 (a) Image of stilus tablet with illumination at 0°. Words visible: QUINQUE, NUTRIUI. (b) Different types of feature detected by phase congruency. Yellow: highlight; blue: transition from highlight to shadow; green: shadow.

(a)

(b)

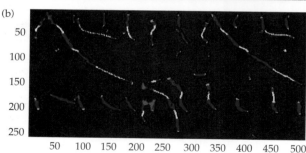

Figure 2.14 (a) The result of detecting candidate incisions reveals gaps in the strokes, mostly caused by the relatively dense woodgrains. (b) The result of applying an algorithm to fill in such gaps (yellow).

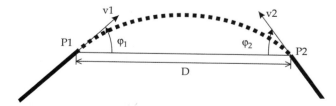

Figure 2.15 The geometry of filling the gap between two stroke portions.

Only two accounts exist which attempt to describe the processes that papyrologists go through when transcribing and translating a text,[12] both demonstrating the complex, recursive nature of the task, the fact that it is a difficult one to describe and communicate, and ultimately, that it is difficult to explain in the final annotated version of the text how the papyrologist got from papyri to translation. From a cognitive psychology stance, although the processes involved in reading have been widely studied, few conclusions have been drawn as to how a reader would approach such damaged, fragmentary, foreign language texts and construct a logical, acceptable meaning.

The problem with trying to discover the process that papyrologists go through while reading an ancient text is that experts are notoriously bad at describing what they are expert at.[13] Experts utilise and develop many skills which become automated, and so they are increasingly unable to explain their behaviour, resulting in the troublesome 'knowledge engineering paradox': the more competent domain-experts become, the less able they are to describe the knowledge they use to solve problems. Against this background of qualitative, descriptive material, we undertook a novel programme of knowledge elicitation, based on a set of techniques borrowed from cognitive psychology and artificial intelligence to aid in the understanding of expert processes in such cases. These techniques are varied to ascertain as much information as possible about the domain, and in this case involved building a knowledge library by undertaking a content analysis of the critical apparatus of published papyri; psychology experiments involving setting individual experts tasks, and comparing and contrasting their techniques; the transcription and analysis of discussions regarding the texts; and structured and semi-structured interviews of the experts involved. Figure 2.16 shows the findings from the analysis of one of these discussions. A transcript of an (approximately) hour-long discussion regarding a small portion of tablet 974 was made, and divided into the salient arguments made: discussion regarding the overall meaning of a document (level 6), the meaning of individual units (5), elements of grammar (4), and the identification of words (3), characters (2), and features (1).

The papyrologists repeat the process of looking at features, letters, and words before assigning meaning to them, and it can be seen that this is a thoroughly recursive process, where sense, meaning of words, and overall meaning of the text are constantly debated. An analysis of this type indicates how knowledge-elicitation techniques can be fruitfully used to gain an understanding of an expert process.

Data regarding linguistics were assembled from the published set of Vindolanda ink tablets[14] — the only set of contemporaneous documents with similar enough provenance and linguistic characteristics to be

[12] Youtie (1963, 1966).
[13] McGraw and Harbison-Briggs (1989).
[14] Bowman and Thomas (1994).

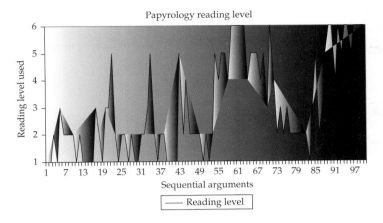

Papyrology reading level

Figure 2.16 Analysis of the different 'reading levels' used by the papyrologists as they attempt to transcribe a text.

comparable to the Vindolanda stilus tablet texts. These texts were marked up and analysed using techniques derived from Corpus Linguistics to provide lexicostatistics (bigram and trigram analysis, word lists, collocations, and so on). Further knowledge elicitation was undertaken with the experts on a palaeographical level, to get them to explain thoroughly the characteristics of individual letters, the way in which individual strokes are made, and how they combine to form characters.

The findings from this research were related to current cognitive psychology theories regarding the resolution of ambiguity in texts during the process of reading. This is a phenomenally complex process regarding which there are many papers, but few theories actually implemented as computer programs to test the models postulated. Parallels between one of these major theories, the interaction activation model of visual word recognition, which aims to explain the word superiority effect (the fact that a word can be identified more accurately than a single letter),[15] and the findings of the knowledge-elicitation experiments were obvious.

Word and Character Corpus

The basis of the computer model of the process of interpreting a stilus tablet that we have developed during the project and with which a historian could interact is a *corpus* of words and character shapes. The word corpus should provide *inter alia* the prior probability of any word that seems a candidate interpretation of a region of the text, and expected variations in the shapes of characters that are expected to appear, as well as information such as bigram and trigram frequencies of pairs and triples of characters.

Contemporary Latin word corpora have been published previously, most notably the 5 million words that appear in the Perseus database. From the perspective of historians, the frequency of occurrences of words in the Perseus database is not necessarily a good predictor either of the word corpus or word frequencies in an analogous Vindolanda database. Not least among the reasons for this is the fact that Vindolanda represents one of the most northern, isolated settlements of the

Roman empire, and it was staffed largely by mercenaries recruited from modern northern Germany and Belgium.

In collaboration with Bowman, we have developed a Vindolanda corpus of words and character shapes (and variations). The word database was constructed from the ink tablet texts from Vindolanda that had previously been interpreted by Bowman and Thomas (1994). The frequency histogram of the 1500 characters appearing in those documents matches closely the character frequency histogram from the Perseus database. The word and character frequencies are used to compute likelihoods of interpretations (and the description length estimation) in the reading system described in the next section.

The character shape models were derived from overlays written over the top of the document by Bowman. These overlays were then dissected into: strokes (each with a length, orientation, and curvature descriptor), end points (with an associated type, e.g. serif), and junctions (each with a type, relative orientation, etc.). The character descriptions were then compiled into SGML format. Compiling the character database has been a major undertaking. The painstaking process analysed 8 characters per hour. A typical six-hour stint during a day produced 48 such characters. With checking, the total time taken was 40 days equivalent, spread over a two-month period. However, as noted in the introduction, the resulting character shape database, and the software and descriptors developed for its production, is likely to prove a major resource for palaeography research for decades to come. Its accessibility to the scholarly world is being enhanced by writing Java routines to scan the SGML and provide Internet browsing.

Reading the Texts

We have made sufficient progress to begin to put the pieces together in the form of a computer model of the process of interpreting (or helping a historian to interpret) a stilus (or ink) tablet, utilising the findings and the statistics from the knowledge-elicitation experiments outlined above. This software uses a multi-layered reflective interpreter to segment, label, and parse images

[15] McClelland and Rumelhart (1986).

to produce an image description the way that a human would,[16] utilising the type of semantic information the linguistic and palaeographical knowledge-elicitation techniques have revealed. This is not an attempt to build an 'expert system' that will automatically 'read' and provide a transcription of the texts; rather it is a means by which papyrologists will be able to mobilise disparate knowledge structures, such as linguistic and visual clues, and use these in the prediction process to aid in the resolution of the ambiguity of the texts. The advantage of developing such a system is that it enables papyrologists to maintain an explicit record of the alternative hypotheses developed, and to switch effortlessly between these initially competing hypotheses, allowing them to see the development of their reading of the texts and trace any conclusions back to their initial thought processes. The overall aim is to aid in the transcription of the stilus tablets, but it is hoped that the tools may also be used by papyrologists working on other texts, and through this process we are gaining a further insight into the papyrology process itself which is relevant to researchers working in cognitive psychology.

It should be noted that the representations of character shapes used in the current preliminary implementation are considerably impoverished relative to what can now be delivered by the image analysis algorithms that we have developed. Elaborating the reading process with such enhanced models is further work to be undertaken. In essence, characters are normalised in size and aspect ratio, so that they can be represented by the end-points of their strokes. The set of character shapes from the corpus was divided at random into a 75 per cent set used for learning the character models, and the remaining 25 per cent for testing it. A novel clustering algorithm[17] was used to identify the cloud of points corresponding to each stroke end-point in the training set, each cluster was fitted with a 2D Gaussian, and this in turn was used to assign to each stroke end-point in each subsequent test image a prior probability that it was (say) the top end of the top stroke of an 's'. The prior probability of words was computed from the word corpus, as were the bigram frequencies of characters. The levels used in the reading program were character, bigram, and word. Character identities (and so on) were chosen at random in a Monte Carlo fashion, and at each iteration a description length computed (see Robertson and Brady, forthcoming, for details). The minimum description length interpretation was finally chosen. In general this took about fifteen iterations. The following shows the iteration process converging to the correct interpretation of a line of Latin text.

First, the entire text (*Tab. Vindol.* II 255) is shown in its original form, to illustrate the complexity of the problem (Figure 2.17(a)). A zoom on a portion of the text is shown in Figure 2.17(b), and the strokes detected in this image in Figure 2.17(c). A snapshot of the screen during the interpretation process, leading to the MDL (minimum description length) interpretation of the text, is shown in Figure 2.17(d).

The final interpretation is: **ussibuss puerorum meorum**, which (amazingly) is correct!

Conclusions

Stilus tablets pose difficult image analysis challenges. To amplify the shallow incisions, we have developed a novel technique which we call *shadow stereo*. The algorithm itself is straightforward: the devil is in the detail, in particular in the detection of the features of interest, specifically shadows lying adjacent to highlights. The work has progressed to the point at which we are ready to investigate palaeographic models of character formation in order to fill the gap between what the image analysis can deliver and the knowledge level sketched in the previous section. Recent progress in artificial intelligence[18] suggests a way forward. Still, there is much to do. First, we have to date restricted our attention to a single azimuth direction. It is clear that we should be able to combine the results from several azimuths to provide a more complete analysis of the tablet, after the fashion of tomographic reconstruction in medical image analysis. Second, stilus tablets, though immensely difficult to analyse, are easier than other artefacts, for example lead curse tablets. We look forward to years of collaboration between us as engineers and our historian colleagues.

As for future work:

1   There remains much to do to make the technology and resources developed during the project usable by historians who do not possess (and do not want to possess) image processing and Java skills. A significant human–computer interface effort is required to make the system available over the Internet.
2   The computer architecture developed toward the end of the project uses a far more impoverished representation of strokes and characters than the image analysis system can in fact deliver; however, significant effort is required to upgrade the reading program to work with such enriched representations.
3   Equally, the reading program deploys a limited stock of knowledge: character shapes, bigrams (and their prior probabilities), and a corpus of words (and their prior probabilities). It does not, for example, mobilise knowledge of Latin grammar, including expected word endings. Nor does it incorporate knowledge — however scant — about the subject matter of the Vindolanda tablets.
4   The practical challenge of digitising the tablets to high quality (many are in the British Museum) left little time for imaging tablets from different azimuths, hence opening up the way to tomographic analysis of those tablets.

---

[16] Robertson (2001).

[17] Robertson and Brady (forthcoming).

[18] Robertson (2001).

(a)
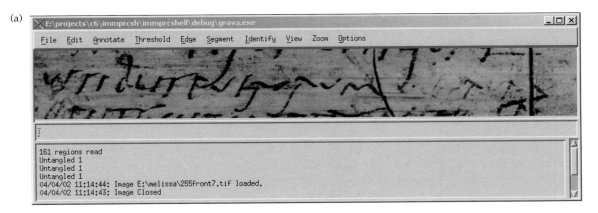

(b)
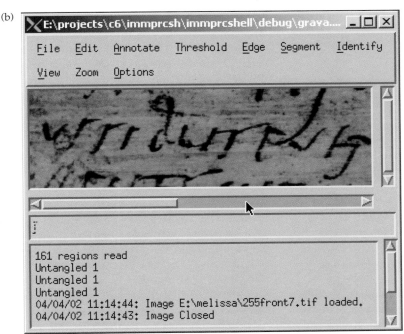

(c)
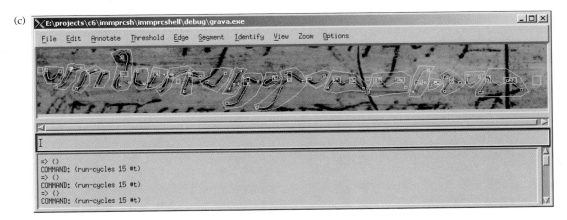

(d)

iteration 0 MDL=86.36351 interpretation= (ussiburr   pierorum meurum)

iteration 1 MDL=56.676826 interpretation= (ussibuss   pierorum meorum)
iteration 2 MDL=56.111915 interpretation= (ussiburr   puerorum meorum)
iteration 6 MDL=47.301033 interpretation= (ussibuss   puerorum meurum)
iteration 7 MDL=36.863136 interpretation= (ussibuss   puerorum meorum)

Figure 2.17 Screenshots of *Tab. Vindol.* II 255. (a) Part. (b) Close-up. (c) Stroke detection. (d) MDL interpretation of text.

5 Finally, since our collaboration began, confocal microscopy has made great strides. This was always the basis of an alternative approach to imaging the stilus tablets, with the advantage that depth is explicit, unlike in shadow stereo, where it is implicit.

Note

This work would not have happened, nor followed the course that it has, without the support, encouragement, and drive of our colleagues Alan Bowman, Charles Crowther, and Roger Tomlin at the Centre for the Study of Ancient Documents. Nor would it have happened without the financial support of the Leverhulme Trust and the EPSRC.

References

BOWMAN, A. K. (1999), 'The Vindolanda writing tablets, 1991–1994', *Atti del XI Congresso Internazionale di Epigrafia Greca e Latina . . . 1997* (Roma): 545–51.

BOWMAN, A. K. (2001), 'Imaging incised documents', *Atti del XXII Congresso Internazionale di Papirologia* (Firenze): 147–50.

BOWMAN, A. K. and DEEGAN, M., eds. (1997), *Imaging Documents, Literary and Linguistic Computing* 12/3.

BOWMAN, A. K. and THOMAS, J. D. (1994), *The Vindolanda Writing-Tablets (Tabulae Vindolandenses II)* (London).

BOWMAN, A. K. and TOMLIN, R. S. O. (2003), 'Wooden stilus tablets from Roman Britain', [Chapter 1 of the present volume].

BOWMAN, A. K., BRADY, J. M. and TOMLIN, R. S. O. (1997), 'Imaging incised documents', in A. K. Bowman and M. Deegan (eds.), *Imaging Documents, Literary and Linguistic Computing* 12/3: 169–76.

CROWTHER, C. V. (1997), 'Imaging inscriptions', in A. K. Bowman and M. Deegan (eds.), *Imaging Documents, Literary and Linguistic Computing* 12/3: 163–7.

FAUGERAS, O. (1993), *Three-Dimensional Computer Vision* (Cambridge, MA, and London).

FORSYTH, D. and ZISSERMAN, A. (1990), 'Shape from shading in the light of mutual illumination', *Image and Vision Computing* 8/1: 42–9.

FORSYTH, D. and ZISSERMAN, A. (1991), 'Reflections on shading', *IEEE Transactions on Pattern Analysis and Machine Intelligence* 13/7: 671–9.

GIBSON, E. J. and LEVIN, H. (1975), *The Psychology of Reading* (Cambridge, MA, and London).

GRANLUND, G. H., KNUTSSON, H., WESTELIUS, C.-J. and WIKLUND, J. (1994), 'Issues in robot vision', *Image and Vision Computing* 12/3: 131–48.

HORN, B. K. P. (1972), 'Understanding image intensities', *Artificial Intelligence* 8/2: 201–31.

HORN, B. K. P. (1986), *Robot Vision* (Cambridge, MA).

KOVESI, P. (1999), 'Image features from phase congruency', *Videre* 1/3 (http://mitpress.mit.edu/e-journals/Videre/).

MALLAT, S. G. (1999), *A Wavelet Tour of Signal Processing* (San Diego and London).

McCLELLAND, J. L. and RUMELHART, D. E. (1986), 'A distributed model of human learning and memory', in D. E. Rumelhart, J. L. McClelland, and T. P. R. Group, *Parallel Distributed Processing*, vol. 2, *Psychological and Biological Models* (Cambridge, MA).

McGRAW, K. L. and HARBISON-BRIGGS, K. (1989), *Knowledge Acquisition: Principles and Guidelines* (London).

MOLTON, N., PAN, X., BRADY, M., et al. (2002), 'Visual enhancement of incised text', *Pattern Recognition* 36/4: 1031–43.

MORRONE, M. C. and OWENS, R. A. (1987), 'Feature detection from local energy', *Pattern Recognition Letters* 6: 303–13.

ROBERTSON, P. A. (2001), 'Self-adaptive architecture for image understanding', D.Phil. thesis (Oxford).

ROBERTSON, P. and BRADY, M. (forthcoming), 'A novel clustering algorithm', *Artificial Intelligence* (submitted).

RUTKOWSKI, W. S. (1979), 'Shape completion', *Computer Graphics and Image Processing* 9: 89–101.

TERRAS, M. (2000a), 'Towards a reading of the Vindolanda stylus tablets: engineering science and the papyrologist', *Human IT* 2/3: 255–71 (http://www.hb.se/bhs/ith/humanit-eng.htm).

TERRAS, M. (2000b) 'Border crossing: engineers, papyrologists, and the user interface', Association of Computing and the Humanities and Association of Literary and Linguistic Computing Joint Conference, University of Glasgow.

TERRAS, M. (2001), 'Reading the papyrologist: building systems to aid the humanities expert', Association of Computing and the Humanities and Association of Literary and Linguistic Computing Joint Conference, New York University.

TOMLIN, R. S. O. (1998), 'Roman manuscripts from Carlisle: the ink-written tablets', *Britannia*, 29: 31–84.

TOMLIN, R. S. O. (1999), 'Curse tablets from Roman Britain', *Atti del XI Congresso Internazionale di Epigrafia Greca e Latina . . . 1997* (Roma): 553–65.

ULLMAN, S. (1976), 'Filling in the gaps: the shape of subjective contours and a model for their generation', *Biological Cybernetics* 25: 1–6.

WILSON, T. (1990), 'Confocal scanning microscopes', *Proceedings of SPIE: the International Society for Optical Engineering* 1319: 460–1.

WOODHAM, R. J. (1980), 'Photometric method for determining surface orientation from multiple images', *Optical Engineering* 19/1: 139–44.

WOODHAM, R. J. (1994), 'Gradient and curvature from the photometric-stereo method, including local confidence estimation', *Journal of the Optical Society of America A* 11/11: 3050–68.

YOUTIE, H. C. (1963), 'The papyrologist: artificer of fact', *Greek, Roman and Byzantine Studies* 4: 19–32.

YOUTIE, H. C. (1966), 'Text and context in transcribing papyri', *Greek, Roman and Byzantine Studies* 7: 251–8.

CHAPTER 3

Digitising Cuneiform Tablets

Carlo Vandecasteele, Luc Van Gool, Karel Van Lerberghe, Johan Van Rompay &
Patrick Wambacq

In 3200 BC an 'intellectual' living in the city of Uruk in what is now south Iraq is said to have invented writing. By using a stilus on a moist clay tablet he managed to transfer his message to 'fellow scribes'. Ancient languages, such as Sumerian, Akkadian, Hittite and many others, were inscribed on tablets from then until the beginning of the Christian era. Approximately 150 years ago, in 1850, 'Assyriology', the science of reading and interpreting cuneiform signs, was created. At that time adventurers travelled the Middle East searching for inscriptions, spending months and even years copying these inscriptions for study in Europe.

The Epigraphist, the 20th-Century Monk

Now, at the beginning of the third millennium AD, not much has changed. Epigraphy is probably the most conservative science at present in existence. The basic tools are those used 150 years ago: the lead pencil, a magnifying glass, and calibrated paper. In 2000 AD the job of the epigraphist is practically the same as that of the Sumerian or Assyrian scribe in 2000 BC, 4000 years ago. Today epigraphists are members of the Middle Eastern archaeological missions, and the cuneiform tablets discovered are still copied by hand. To be successful the scholar has to understand the values and forms of approximately 600 signs, and how the meanings evolved over the 3000 years that cuneiform was in use. A hand copy of a tablet remains highly subjective and depends on the artistic qualities of the copyist (cf. Figure 3.1). The task is tedious and time-consuming, and it is difficult to correctly portray the nuances and ambiguities of an original text. Since Middle Eastern antiquities departments do not allow the export of excavated tablets, more efficient field methods must be developed for registering the tablets. It is essential the new method reproduces the original tablet in a format that conveys all the cuneiforms exactly as inscribed.

The Computer, the 21st-Century Monk's Stilus

Even the monk does not escape advanced technologies. At Middle Eastern sites, philologists and historians require new methods to record quickly and accurately the information contained in excavated tablets.

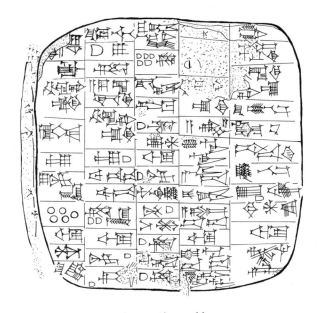

Figure 3.1 Hand copy of a cuneiform tablet.

The First Step: Using Ammonium Chloride on Tablets

The ancient Mesopotamian scribe wrote his messages by pressing a stilus in moist clay tablets. The tablet was then dried or baked. The result is a brownish, convex clay lump with 'depressions' (the signs). These cuneiforms are easily recognised by eye, but very difficult to photograph properly. To improve the contrast between the signs and the clay, Assyriologists traditionally apply ammonium chloride powder to the tablet. When ammonium chloride crystals are placed in a glass pipette and heated by a Bunsen burner, the fumes produced are blown over the tablet. The tablet turns white and the signs are more apparent. This improves the photography; however, the process is time-consuming and produces a very unpleasant odour.

Lately, a more convenient method has been developed. White fumes of ammonium chloride are obtained from the reaction between hydrochloric acid (HCl) and ammonia (NH_3). Two open containers (e.g. drinking glasses) are placed in a vessel — one contains concentrated hydrochloric acid, and the other concentrated ammonia. The vessel is covered by a glass plate with several tablets between the two reagents. In approximately 10 minutes a cloud of ammonium chloride effloresces, covering the tablets in a fine

Figure 3.2 Two cuneiform tablets, one before and one after chemical treatment.

powdery crust. No damage whatsoever is caused to the tablets. Figure 3.2 shows a tablet before and another after treatment. The improved quality and contrast are clearly apparent.

Second: Digitising the Tablets

To obtain a sharp shadow that will be picked up by a camera, a raking light is used that enhances the writing. Such illumination is not uniform, since it comes from a single light source at the left side of the tablet. As a result signs on the left will be too light and those on the right too dark. Furthermore, the tablets are not flat and the cuneiform signs do not have a uniform depth. The resulting photography with its changing focal length leaves some signs less sharp (Figure 3.3(a)).

To overcome these problems the quality of the images is enhanced at the excavation by use of computers. The technique, developed at the ESAT laboratory in Leuven University, consists of traditional photography and scanning the pictures into a computer, or using a digital camera and running an image processing algorithm program to obtain the desired results.

In 1994 this method was initiated in Syria on the Tell Beydar tablets (2400 BC). Photographs of the tablets were taken at Beydar and scanned at the ESAT laboratory. A problem was in the varying picture quality which could not be controlled at Tell Beydar. Therefore, a digital camera was used to register the tablets during the following excavation seasons. An AgfaCam camera on a traditional Minolta body with macro-lenses for recording the details of inscriptions or sealings was chosen. The image is stored on a digital card (80 images possible) for transfer to a laptop. The advantages are obvious: the image quality can be controlled on the site, the image can be given a reference number linking the tablet to the archaeological context, and the image quality can be improved on site by running image processing algorithms.

Processing Images

For the Assyriologist working in Middle Eastern excavations a fast method to adequately register and study tablets is a priority. Hence, the partly unreadable photographic images of the tablets, resulting from non-uniform illumination, are a serious impediment. The unsharp perimeter of a convex tablet is acceptable, if the illumination problem can be solved with digital enhancement techniques. If 95 per cent of the tablet becomes readable, the tablet edges can still be copied by hand.

Image Quality Enhancement

To understand how the non-uniform illumination is compensated, the notion of 'spatial frequency' has to be explained. Spatial frequency (or frequency in the spatial domain) is the speed or frequency at which the grey levels of the pixels in an image vary across the image. Since each image is two-dimensional, the spatial frequency is divided into a horizontal and vertical component. Looking at tablet photographs, two spatial frequencies can be distinguished: a variation in grey levels due to the non-uniform illumination, and a variation of grey levels due to the presence of relief (the impressed signs). The frequency of the first component is much smaller (a smoother variation) than the latter component. Suppressing the smooth variation annihilates the effect of the non-uniform illumination, while the useful information (the writing) is accented.

To separate the low frequency from the high frequency content, a straightforward method was chosen which is easily implemented. The average grey value in every pixel of the image is computed. This is done by summing the grey values of all pixels in a certain neighbourhood of the pixel under consideration. The sum is divided by the number of pixels involved. The neighbourhood size is important: (1) the use of a small neighbourhood (involving a small number of pixels) results in an average value with greater variation of grey pixels; (2) the use of a large neighbourhood (the entire image could be used) gives an average value with less grey variation. This process can be regarded as if a window is slid over the image, whereby in every position of this window all pixels lying inside it are used to calculate an average value, which is attributed to the pixel in the centre of the window. As a result a new image appears which is a filtered version of the original image. When the size of the window is large, a filtered image is obtained that contains only very low frequencies (very smooth variations), while the use of a smaller window provides an image containing higher frequencies. In this way, the created filter allows low frequencies to pass, and catches or suppresses high frequencies. Thus the border between what passes and what is blocked is determined by the size of window used.

To enhance the clay tablet images, the appropriate window size was determined experimentally. It turned out that a size of 148×148 points is appropriate. The result of applying such a filter to the original images can be seen in Figure 3.3(b): the high frequency content (the writing) is filtered out, the remainder being a slower variation, originating from the difference between the foreground and the background and from the non-uniform illumination.

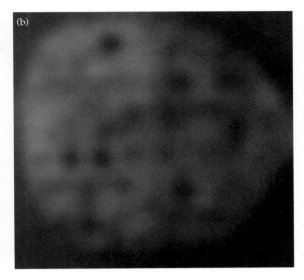

Figure 3.3 (a) Digital photograph of a cuneiform tablet. (b) Filtered version of original tablet. (c) Image after application of unsharp masking with media filter.

When the low frequency information is abstracted from the original image, the result is an image with a global average of zero. Readjustment is made by adding to every pixel value the global average of the original image (a constant value for the entire image). In this way an image is created with the same average value as that of the original image, but containing only the high frequency content (the writing). The effects of the non-uniform illumination have disappeared.

The details on the tablet are enhanced by applying an extra filter (a process called 'unsharp masking'). In some pixels the subtraction, followed by the addition of the global average, may result in negative numbers or numbers greater than 255, which is outside the 8 bits range used to represent the images. This can be corrected by forcing all negative values to zero and all values greater than 255 to 255. Now the image still contains some noise. This noise is suppressed by using a 'median' filter. As a result a much clearer final image is obtained (Figure 3.3(c)).

Conclusion

As stated in the introduction, copying tablets and drawing sealings is done manually by epigraphists and sigillographists in the traditional way. This procedure is still valuable; the contact between the modern scholar and the antique object cannot be overestimated.

If an archive is discovered, the procedure of drawing hundreds of signs and figures by hand is time-consuming and obliges the researchers to work on the site for many months or even years. The publication of new archives is often delayed by years. Researchers working in Middle Eastern excavations, being confronted with such problems, have been seeking new, faster ways to record the scientific data contained in the archaeological material. New techniques of digital imaging developed by engineers in collaboration with philologists and archaeologists offer now the best opportunities to solve these problems.

These digital techniques are now being used continuously by the authors on archaeological sites in the Middle East. Cuneiform tablets are being photographed by digital camera (with traditional lenses) linked to a laptop. At least six images are taken of each tablet: obverse, lower edge, reverse, upper edge, and left and right side. Through the enhancement method described the quality of the image is created. The final clear image is stored on an Iomega Jaz drive and paper copies are printed. The epigraphist now has an excellent working document with legibility of 90 to 95 per cent. The advantages of the digital system are obvious. A great number of tablets can be registered during the excavation season. The images can be given identification numbers linking them to archaeological layers, architectural data, objects, and so on. Once recorded the images are available to scholars around the world.

It is clear that some distortion occurs when switching from a three-dimensional tablet to a two-dimensional image. This distortion does not exceed the distortion created if drawing the tablet by hand. The distortion has no influence on the legibility or interpretation of the text.

The use of digital registration is not limited to cuneiform tablets. All sealings (seals unrolled on clay lumps), pottery sherds, and fingerprints on clay vessels can be registered and imaged in this manner.

At present, hand copies of cuneiform tablets are still being drawn for final publication. The quality of these hand copies has improved considerably by using the digital images. It is clear that in the future all publication of cuneiform tablets will be done digitally. Currently, sign lists are made using the digital records. In the future, tablets will also be interpreted by computer. Several scientific institutions are now developing language programs to achieve this aim.

Note

This paper presents research results of the Belgian programme on Interuniversity Poles of Attraction (V/14).

CHAPTER 4

Interpretation of Ancient Runic Inscriptions by Laser Scanning

Jan O. H. Swantesson & Helmer Gustavson

Introduction

Investigations of runic inscriptions have traditionally been done in the field by eye and by feeling the outlines of runes with the fingers. Photographs taken from different angles have been of complementary use. The modern laser scanning technique has proved to be useful in many cases where the interpretation otherwise is uncertain. It records the height of surfaces in shades of grey only, so that disturbances caused by different colours of the minerals in the rock disappear. The article describes briefly the laser scanner used for the purpose and how data from its measurements can be treated. An account is given of runic inscriptions in Scandinavia where the technique has been applied. It is stated how and under which circumstances significant new information compared with traditional methods of interpretation can be gained. The methods that can be used when interpreting runic inscriptions with the new technique are also described. Swantesson has written the parts about the technique and how the measurements are performed, while Gustavson is responsible for the interpretation of the rune readings.

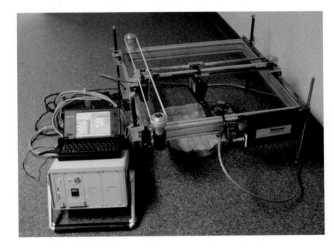

Figure 4.1 Photograph of the complete laser scanner equipment in the laboratory. Measurements of a piece of limestone from the province of Gotland are being performed.

The Laser Scanner

The current version of the laser scanner is reproduced in Figure 4.1. A commercially available triangulation laser gauge probe takes height measurements. Its low power GeAs laser emits a pulsed beam of red light with a wavelength of 670 nm. Where the beam hits a surface it creates a spot with a diameter of about 0.2 mm. The reflected image of this spot is detected by a position-sensitive light sensor mounted alongside and at an angle to the laser source. By this method the laser probe used can obtain height values with a resolution of 0.025 mm and an approximate accuracy of 0.1 mm within a measurement range of 100 mm.

A flat bed scanner, where belts driven by electric stepping motors move the probe in the *x*- and *y*-directions, has been constructed. The four legs are adjustable in height, and usually the best results are gained when the scanner has approximately the same inclination as the surface to be measured. On steeply inclined objects the device has to be tightened to them by straps. The maximum area that can be scanned without having to move

the scanner is 400 × 420 mm. Two rechargeable motorcycle batteries power the entire system, which works on 24 Volts.

Control boards for the motors as well as the laser are contained in a separate box. A portable computer collects the measurement data. Special software has been written for the data acquisition process and for control of the motor movements. It is possible, for example, to choose the size of the area to be measured and its origin. It is also feasible to choose the distance between height values in the *x*- and *y*-directions that are saved in the computer. The minimum interval is 0.4 mm in both directions. For convenience a distance of 1 mm, making one height value for each square millimetre, has been used in the majority of the measurements performed.

When constructing the laser scanner equipment much effort has been made to keep it as light as possible, thereby facilitating its handling in the field. The total weight of all the parts together is about 30 kg. Two people can therefore carry it some distance fairly easily. The machinery can be transported without any trouble in almost any make of car. A complete measurement over an area sized 400 × 400 mm, with one height value for every square millimetre saved in the computer (a total of 160,000 height values), takes approximately two hours to perform. A more thorough technical description of the laser scanner, as well as comparisons with other methods for detailed measurements in the field, is found in articles by Swantesson[1] and Williams et al.[2]

The data are available for further treatment immediately after the scanning is completed. They are stored as an array of height values where the distance in both the *x*- and the *y*-direction is known. Statistical calculations can be performed of, for example, surface roughness.[3] By the use of geodetic and GIS (geographic information systems) software packages the results of the measurements can be presented in various ways. Contour maps, 3D models, and digital shadow images are only some of the possibilities useful when interpreting a scanned surface. The digital shadow images resemble ordinary photographs. The relief is, however, reproduced by having an imaginary sun shining from a direction of your own choice. Neither is the picture obscured by different colours or shades of grey on the original surface.

Applications of the Laser Scanning Technique

A prototype of the laser scanner was constructed as early as the late 1980s. At that time all control units were placed in the back of a van and the scanner was connected to them by 30 m long cables. A petrol-driven alternating current generator powered the device. The total weight of the entire machinery was well over 100 kg. The aim of these early measurements was to get an objective measure of the form of minor weathering features on different types of rocks in Sweden.[4] The machinery has since then been gradually improved.

In 1994 a still ongoing project concerning breakdown and weathering of Swedish Bronze Age rock carvings started with financial support from the Swedish National Heritage Board.[5] The state of about twenty-five rock carvings in various rock types and surroundings in southern and central Sweden is monitored every or every second year. One of the main objectives of the project is to calculate the loss of material from the sites investigated. This is done by repeated measurements. In order to be able to superimpose micro-mappings performed at different times, minor studs are drilled into the rock in the immediate vicinity of the carvings. The studs serve as fix points. Results up to now indicate that practically no changes are observable on about half of the rock carvings. In some places, however, severe loss of material, amounting to more than a cubic centimetre from a measured area of about 300×300 mm, has occurred within a few years. This deterioration will of course obscure the outline of the rock carvings.

In 1999 an EU project on erosion and downwearing on European rocky coasts commenced. Natural processes along coastal stretches of the Mediterranean, the Atlantic, the English Channel, and the Baltic are compared to one another.[6] To study the course of micro-erosion three laser scanners of the type described in this article are used in the different countries participating in the project.

When measurements of runic inscriptions on stones were made it was discovered that several details that were not readily seen with the naked eye were revealed on images produced by laser scanning. For this reason the method has been used as a tool for interpreting some inscriptions where the meaning of parts of the text is somewhat doubtful. The recognised inconvenience in using a triangulation laser with undesirable reflections and parts that are not recorded because of abrupt height changes on the measured surface is not important with objects of this type. The micro-relief of the carvings is generally fairly smooth. Other scientists have also used laser scanner techniques for studying the shape of grooves on rune stones.[7] This makes it possible to distinguish between individual carvers. A slower laser scanner, but with a better resolution in all directions, is used. The measured areas are considerably smaller than when text interpretation is made.

Interpretation of Runic Inscriptions

Rune stones are fairly common in Scandinavia, and several thousands are known. The majority of them are found in the province of Uppland, just north of Stockholm. There are two basic forms of the futhark, the runic alphabet. The so-called older futhark contains twenty-four characters and was in common use from the second century up to the eighth century. As a result of rapid language changes the new futhark of the Viking Age, containing only sixteen characters, developed. The interpretation of runes made with the older futhark can often be problematic. The vocabulary differs considerably from later texts, and the relatively few inscriptions left are short. The limited number of runes in the new futhark was not sufficient to reproduce all phonemes in the language. The rune for 'k' can, for example, also mean 'g', and the rune for 'u' can equally well be read as 'v' or 'y'. This means that a single sign in some cases may be misinterpreted and lead to a reading that is not in accordance with the actual meaning of the text. A further complication is local variation of some runes.

When interpreting inscriptions made by the runes of both the old and the new futhark it is important that the readings are as certain as possible. It is, for example, vital when studying the language of the older runes and also essential for a better understanding of runic texts in general. The runic inscriptions that have been micro-mapped with the laser scanner and are described in this article are listed in Table 4.1 and their locations are seen in Figure 4.2.

---

[1] Swantesson (1994), 209–13.
[2] Williams et al. (2000a).
[3] Swantesson (1992a) and (1994), 212–19.
[4] Swantesson (1989), 131–47, 184–93, and (1992b), 277–8.
[5] Swantesson (1996).

[6] Williams et al. (2000b).
[7] Kitzler (2000).

Table 4.1 Locations of runic inscriptions

Designation, site and province[a]	A[b]	B[c]
1. U 287 Vik, Hammarby 106, Uppland	200 × 300	60,501
2. 14342, Lokrume, Gotland	75 × 940	284,031
3. Östra Eneby 1 (Himmelstalund), Östergötland	190 × 650	121,047
4. NF 1929, Östra Husby 252, Östergötland	315 × 330	115,923
5. Ög 136, the Rök stone, Rök, Östergötland	250 × 290	73,041
6. DR 345, Simris 2, Simris church, Skåne	300 × 300	90,601
7. 29160, Bo, Tanum parish, Röö (Otterön), Bohuslän	380 × 570	217,551
8. NF 1992, the Skramle stone, Gunnarskog, Värmland	135 × 810	158,599
9. N-449, the Kuli stone, Trondheim, Norway	140 × 1,130	228,551

[a] The numbers preceding the different objects correspond to the numbers in Figure 4.2 showing their location.
[b] Size of the mapped area (mm).
[c] Number of height values registered during the measurements.

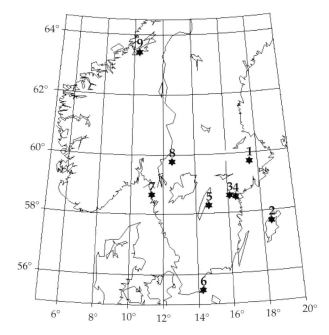

Figure 4.2 Sketch map of the locations of the runic inscriptions described in the article. The locations are listed in Table 4.1, with the same numbers as on the map.

Scanning of Rune Stones for Documentation Purposes

The aim of micro-mapping the rune stones at Vik, Rök, and Simris (nos. 1, 5, and 6 in Table 4.1 and Figure 4.2) has not been primarily to interpret the text. Instead a detailed documentation of the micro-relief of parts of the stones was wanted. The measurements can then be compared to later investigations to see whether any weathering or other damage has occurred. The stone at Vik has been documented continuously for several hundred years. First drawings were made and later photographs taken. Both the front and the back of the stone are very smooth. It seems that these surfaces follow the structure of the gneiss. Damage due to flaking, where a superficial layer measuring 3 to 10 mm has been lost, can be observed in various places. From older documentation it can be seen

that this damage is growing. The laser scanning includes parts where flaking has taken place. There are only a few signs of other types of weathering.

The stone at Rök has the longest and also the most famous runic inscription in Sweden. It stands on a gravel bed and is protected from rain and snow by a roof. The rock is for the most part in good condition and the runes are usually easily distinguishable. The micro-mapping took place over one of the lowest rows with runes and a part beneath it without inscriptions. A few damaged areas resembling flaking can be observed below the runes. They do not, however, seem to be active at present.

At Simris two rune stones of quartzitic sandstone are situated a few metres from a road. The smaller stone, over which the measurement took place, is fairly heavily weathered. In many cases the fine-grained mortar used for mending can be detached with a finger. There are also a few loose fragments. The larger stone is in a much better condition and microforms due to glacial erosion are still preserved on it.

The Stone from Lokrume, Gotland

The stone is carved with runic inscriptions as well as pictures (no. 2 in Table 4.1 and Figure 4.2). Only the rows with runes have been measured. The grooves of the runes are very narrow and seem in several cases to be filled with some type of mortar. The rock is limestone, and large parts show signs of intense wearing. In many places the runes are completely gone. On such smooth, even surfaces the laser scanning technique does not add any new details facilitating the interpretation of the text. The probably accidental filling with mortar in some places has caused the original relief to disappear, and runes here are of course not detected by a device measuring heights accurately.

The Inscription at Himmelstalund, Norrköping

In the Himmelstalund Park (no. 3 in Table 4.1 and Figure 4.2) in the town of Norrköping, some 50 surfaces with Bronze Age rock carvings of more than 1000 figures are present. There is also one inscription having only

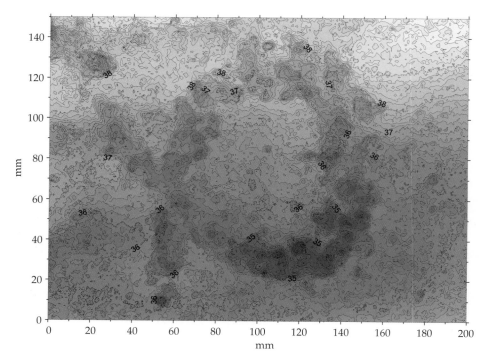

Figure 4.3 Contour map, with a contour interval of 0.2 mm, of the last rune of the inscription **braido** at Himmelstalund, Norrköping. The contours are complemented by darker shades of grey on deeper parts of the rock. The rune is considerably twisted compared to the other runes in the inscription. The cutting was very roughly made, and still preserved large, uneven cut marks are easily seen on the image.

six runes of the older futhark on a rock outcrop. At Himmelstalund large flat rock areas, mainly of mica schist alternating with grass fields, dominate the landscape. The inscription is very negligently carved. The size of the runes varies and they are turned unsymmetrically relative to one another. The bottom of the runes is uneven as well, as seen in Figure 4.3. A first impression of the text is that it is a late counterfeit, especially as the cutting looks very fresh. However, many of the much older rock carvings are very well preserved in the prevailing rock type at Himmelstalund. Another fact that speaks against the text being a counterfeit is that it was discovered in 1871, a time when hardly anybody in the Norrköping area knew about the older futhark. The inscription has been deciphered with conventional methods as **braido**.[8] The reading of **r**, **i**, and **o** is, however, considered doubtful. By micro-mapping the interpretation of the inscription as the female name **braido** can be confirmed with a high degree of certainty.

The Inscription at Oklunda, Östra Husby

The Oklunda runic inscription is cut in an outcrop of a metamorphic rock where amphibole and plagioclase are the dominating minerals (no. 4 in Table 4.1 and Figure 4.2). The carving consists of four 24 cm long horizontal rows with a combined height of 18 cm, and one vertical row with a length of 28 cm and a height of 5 cm. The number of runic characters is about eighty. An argument that the vertical row was engraved later than the horizontal ones cannot be proved. The amphibolite is here somewhat more coarse-grained, which might explain the slightly wider grooves of the runes in the vertical row.

The inscription is a legal document from times before Sweden was Christianised. The form of the runes and the language dates it approximately to the beginning of the ninth century. Since documents of this kind are rare, a correct interpretation of the text is invaluable. There are some difficulties in reading the inscription because the carving is not cut on an ideal surface. Quartz veins protrude above the rest of the surface because of their resistance to weathering, and fissures already existed when the carving was made. There are also some scars due to later deterioration. The reading of the text made with the aid of laser scanning and transcribed into Latin lettering is:[9] **kunar ⋮ faþirunaʀ þisaʀ ⋮ insaflausakiʀ ⋮ sutiuiþita ⋮ insafl* (i)nruþþan ⋮ insabat uifin ⋮ þitta ⋮ faþi ⋮**. An asterisk (*), here and in the following transcriptions, means that there is a rune at this place, but it is impossible to read. Brackets are used when the reading of a rune is uncertain. The letter **þ** corresponds to the English 'th', and the **ʀ** means palatal 'r'; the short vertical lines are used as a punctuation mark. In English translation the rune sentences mean: 'Gunnar engraved these runes, and he fled being due to be punished. He resorted to this sanctuary, and he fled into this clearing, and he effected reconciliation. Vifinn engraved this (as a confirmation).'

Although the majority of the runes are read similarly as in older interpretations,[10] the micro-mapping has revealed several details that make the new reading more reliable. Only a few examples are mentioned here. Salberger[11] arrived at another reading order since he did not detect every occurrence of the punctuation mark ⋮.

---

[8] Krause (1966), 121–3.

[9] Gustavson (2003).

[10] Nordén (1931); von Friesen (1933), 152–3; Salberger (1980), 1–23.

[11] Salberger (1980), 17–23.

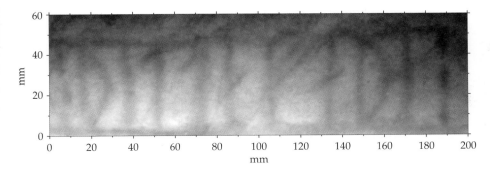

Figure 4.4 Part of the vertical row from the inscription at Oklunda. On the image darker shades of grey represent the engraved runes. The reproduced part reads: ⋮ þitta faþi ⋮, or in English the two words 'this' and 'wrote'. The punctuation marks (⋮), with three short vertical lines, are obvious both at the beginning and the end of the picture.

As seen in Figure 4.4 it is obviously there. The reading of rune no. 48 (after the letter combination **insaf**) as an R by von Friesen[12] is impossible to confirm. There is no sign of it on the contour map produced from the laser scanning. A tiny quartz vein parallel to the structure of the rock in this place might have caused confusion. Neither is it possible to read runes 47–9 as **fik**, as has been done by Nordén.[13]

The Stone at Röö, Tanum, Bohuslän

The Röö stone from Bohuslän, carved with the runes of the older futhark, has been described in great detail[14] (no. 7 in Table 4.1 and Figure 4.2). It is 2 m high and consists of gneiss with a 5 to 6 cm wide pegmatite vein at a height of about 1 m. Pegmatite is a coarse-grained rock type with similar chemical composition to granite. This part has a much rougher surface than the gneiss. The stone exhibits several areas of damage due to exfoliation. It is, however, difficult to determine whether they existed before the engraving was made or if they were caused by later weathering. Another possibility is that the actual cutting of the runes caused the damage. Any present flaking cannot be observed.

The micro-mapping was performed over the part of the stone containing the pegmatite vein. No runes other than in the older traditional interpretations could be found. If runes had existed on the areas lost by flaking they would have been impossible to detect since the depth of the engravings is considerably less than the thickness of the exfoliation sheets. Some minor differences in the form of certain runes compared with previous readings could, however, be observed. For example, it was possible to follow the grooves of the second-last letter (**d**) on row 4 better than was done by Krause[15] when results from the micro-mapping were viewed. A part of the Röö stone where the darkest shades of grey represent the highest inclinations of the rock surface is reproduced in Figure 4.5. The outlines of the edges of areas where exfoliation has occurred appear clearly. The measurements with the laser scanner confirm the former interpretation with a high degree of certainty.

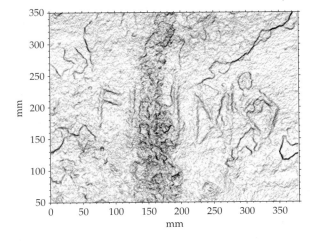

Figure 4.5 Image of a part of the Röö stone, where the degree of darkness is proportional to the inclinations on the original surface. By this type of reproduction the edges where a surface layer has been lost by flaking are clearly revealed. The runes, from the older futhark, are **fahido**, which in English means 'wrote'. The staff of the **f**-rune is worn away and only the twigs remain as two shallow grooves. The **h**-rune is mainly engraved where the surface is rougher because of the presence of a pegmatite vein.

The Skramle Stone, Värmland

The Skramle stone (no. 8 in Table 4.1 and Figure 4.2) was found in 1993 during an archaeological excavation in the parish of Gunnarskog in the county of Värmland. It was one of the foundation stones of a medieval cow-house. The rock type is gneiss and most surfaces are fairly rough and show signs of weathering. Through the fifteen runes of the text there is a vein of potassium feldspar. This vein protrudes somewhat in the micro-relief as well as other quartz and feldspar areas. There are also several fissures on the stone. The carving is made on the natural rock surface. This means that it has not been smoothed in any way before engraving. It is extremely difficult to read the entire inscription with the naked eye. The reason for this is the sharp colour contrasts between the different layers of the gneiss and the weathering the stone has suffered.

The text is written with the runes of the older futhark and it should be read from right to left. The length of the text is 70 cm and the height of the runes varies between 6 and 10 cm. Lines have been cut to frame the upper and lower parts of the lettering. Part of the stone is reproduced in Figure 4.6. In the interpretation of the

12 Von Friesen (1933), 152–3.
13 Nordén (1931).
14 Von Friesen (1924).
15 Krause (1966), 167–70.

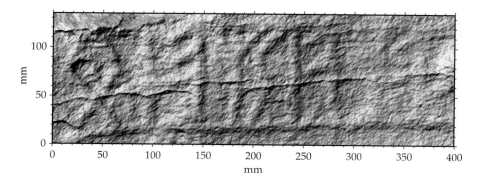

Figure 4.6 Digital shadow image of the second half of the inscription on the Skramle stone. The imaginary light comes from the lower right corner of the picture. The runes, of the older futhark, are read from right to left: **farkaio**. For the reading to be certain it is not possible to view a single digital shadow image only. Structures in the same direction as the light almost tend to disappear, whereas structures perpendicular to the light direction might be exaggerated.

inscription by Gustavson[16] results from micro-mappings were essential. The following sequence of runes was revealed when they were transcribed to the Latin alphabet: ****(j)þaah^ar (f)arkaio**. Only remnants of the first two runes remain and they cannot be interpreted. There might also have existed at least one more rune before the first one. The reading of **j** is somewhat doubtful, but, according to results from field studies, photographs, and the micro-mapping, a **j**-rune is, however, the most likely conclusion. The sequence **h^a** means that the two characters **h** and **a** are combined with each other, forming one single rune. Between **r** and **f** there is a pit with a diameter of 5 cm and a depth of 1 cm. It probably already existed when the runes were made. The pit has signs resembling brackets on both sides. They are purposely made, perhaps for ornamental reasons. The interpretation of the following sign as **f** is somewhat uncertain, while the rest of the inscription causes fewer problems. Although the reading with the help of the laser scanner can be said to have been successful it is not possible to understand what the text means.

Odenstedt has published an alternative reading.[17] It differs in many ways from the interpretation presented here and is not consistent with the results obtained by the micro-mapping. For example, he omits the two first runes, which are almost worn away, and reads the third rune as **o**. From what can be seen on the images produced with the help of the laser scanning technique this is not possible. The most plausible interpretation is as stated above: **j**.

The Kuli Stone, Trondheim, Norway

The Kuli stone (no. 9 in Table 4.1 and Figure 4.2) was rediscovered in the 1950s and is considered to be one of the most important runic texts in Norway, since it is a document contemporary with the earliest days of Christianity in the country. The rock type is gneiss and the inscription is carved on one of the sides of the stone. It starts about 80 cm above the ground and has a length of 110 cm. The combined height of the two rows is never more than 14 cm. They are applied in the same direction as the structures of the rock. Fissures that have been used

as supporting lines for the carving are also present in this direction.

The staves of the runes are perpendicular to the structure, which means that the twigs are cut at an angle to it. At present it is often difficult to observe the twigs clearly. Since they were more difficult to engrave than the staves their grooves might have been made shallower when carving. The surface of the rock is fairly rough, and damaged areas exist along the upper part of the uppermost row. This makes it difficult to follow some of the runes in their full original length. The beginning of the two rows is relatively easy to read. The interpretation difficulties gradually increase towards the end of the rows because the signs are much more worn here. The runes of the stone are painted in red and the item is now housed in a museum. Painting might be good for the public, but it is a hindrance when attempts at new interpretations are made. Since it has a certain thickness the already faint relief is further reduced, making, for example, results from micro-mappings less clear.

Hagland[18] publishes a complete interpretation of the text on the Kuli stone where he makes comparisons with, among other things, a reading from 1956 by Aslak Liestøl. Use of the laser scanning technique has made it possible to confirm large parts of the older reading with great certainty. For other parts an alternative interpretation, however, seems more plausible. The first row is transcribed as:[19] **þurir : auk • halua (:) rþr : rais(t)u • stain : þ(a)nsi • a(f)(t)*******. The signs (:) and (•) are punctuation marks. A translation into English reads: 'Tore and Hallvard erected this stone in memory of *****'. The real number of runes in the row is only forty and not forty-two as has been stated by earlier interpreters. The last five of them can be seen only as five different staves where remnants of twigs are impossible to identify. Older interpretations of the end of the row as **ulfliut** or **ulfrauþ** are for this reason not justified. Another important find is that the three **s**-runes are long-branched, meaning that they fill out the same height as the other signs. In earlier readings they were identified as a short twig equivalent, only having about half this height.

---

[16] Gustavson (1996).
[17] Odenstedt (1997).

[18] Hagland (1998). All the readings of the runes and interpretations of the text concerning the Kuli stone are derived from this article.
[19] Hagland (1998), 130–2.

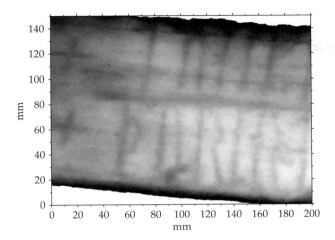

Figure 4.7 Image of the beginning of the two rows on the Kuli stone, Trondheim. Darker shades of grey represent deeper parts. Both rows start with a cross (+). On the first (lower) row the name þurir and the first rune a of the next word are read, and on the second (upper) row tuhlf ('twelve'). The very short twigs on the third rune were not recognised to form a cross in older interpretations. As they are placed slightly to the left of the staff a dotting of the rune i might have been intended, instead of engraving an h-rune.

The second row is read:[20] **tuhlf • uin(t)r • h(a)(f) þi: kr(i)(s)(t)in • tumr • um (•) ri(t)(i)n(u) (:) riki**. In modern English this means: 'Twelve winters had Christianity improved things in Norway'. It can also be translated into the more legal sounding sentence: 'Twelve winters had Christianity secured law and order in Norway'. Seven of the runes of the most central word in the row, **kristintumr** ('Christianity'), were in older readings considered to be uncertain. Now there is no great doubt about the existence of the word. As seen in Figure 4.7, the third rune has the form of a medieval h-rune with very short twigs. Earlier it was read as a short twig **a**. Another possible way to interpret the rune is that the minor twigs were intended to be a dotting of **i**, thus changing the phoneme into **e**. The most important result of the new reading concerns the sequence **um•rit**, runes 27–31 in the second row. Previously it was generally accepted to read **uirit** ('been') here. The new meaning would instead be: '(had) improved/corrected (things)' or 'secured law and order'.

Although uncertainties still exist concerning some of the runes in the inscription, the number of characters where interpretation is not clear has diminished by the use of the laser scanner. The new reading increases the source value of the Kuli stone as contemporary propaganda and maybe also as a missionary for the new Christian belief.

The Laser Scanner as a Tool when Reading Runic Inscriptions

The investigation of runic inscriptions is a philological and epigraphic science. The reading of the text has to

be as certain as possible to yield an interpretation based on it of high quality. Modern techniques such as, for example, laser scanning can add information about the primary source that is not readily observed by traditional methods. They provide the interpreter with a better basis for his or her deciphering of the meaning of the text. This can be compared to a medical doctor whose diagnoses can be improved when modern technical equipment is used for the examination of patients.

A reading of the inscription in the field with the naked eye and by feeling the outlines of the runes on the stone is essential even when micro-mapping by laser scanning is done. An opinion about the composition, rock structures, fissure pattern, and the state of weathering by a skilled earth scientist is also valuable, as there is a risk that rock characteristics can influence a proper reading. The areas where micro-mapping is considered necessary have also to be chosen in the field. If the interpreter is not doing the mapping himself this choice must be done together with the person performing the measurements. At the same time the resolution needed in the x- and y-directions must be decided.

The results from the laser scans are studied with the help of a computer. A first control can be made in the field directly after the measurements have been taken with the computer belonging to the laser equipment. The first step is to study digital shadow images from different directions of imaginary lighting. They resemble photographs taken when the surface is illuminated by a light source from a very sharp angle. Since the images are products of the laser scanning they are not disturbed by, for example, different mineral grains and veins of the rock having various colours. The runes seen on the images are compared with the reading taken in the field.

As soon as there is the slightest uncertainty of how to read a single rune or combination of runes, that part of the text is examined in more detail. Contour maps with a contour interval down to 0.1 mm, 3D models, or images where different shades of grey are proportional to the depth of engraving are then analysed. It is also possible, for example, to produce maps where the gaps between various height levels are filled by contrasting colours, and images where different hues represent varying inclinations of the original surface. In these pictures the 3D information gained by the laser scanning is used to its full extent. Figures 4.3 to 4.7 show typical images used when interpreting texts on rune stones. Paper copies can be printed whenever wanted, notes can be made on them, and different pictures can be compared to one another. The reading of a rune that was difficult to see in the field can in many cases be ascertained by this method. In other cases a few alternative ways of reading can be proposed. The result can also be the exclusion of some interpretations. Another possible conclusion after studying the products of the micro-mappings is that the stone has to be examined further in the field.

It is often useful if two people are doing the reading at the computer together. One of them must be epigraphically skilled and have a thorough knowledge of the language used at the time when the runes were engraved. The other person must master the measurement technique and how to produce the most useful images derived from it. Where texts engraved in stone are concerned, a knowledge of earth science, enabling the evaluation of the surface on which the inscription was made, is also of great importance. Readings made with the help of laser scanning have in many cases proved to be more certain than when only traditional methods are used and they thus provide a better source material. Since the interpretation of a text evidently is based on its source, the applied method will help us to understand ancient runic inscriptions better.

Conclusions

The laser scanner described in this article has fulfilled most of the requirements that can be demanded of a field instrument of this type. It is portable, the cost of the entire machinery is less than £20,000, and it is fairly easy to use. The resolution and accuracy of measurements are also excellent for the applications where the device has been used. It has been possible to perform micro-mappings that could not have been done by purely mechanical methods. Although the speed is fast enough for the present use of the laser scanner, it is still too slow for some other possible applications.

When used on items belonging to our cultural heritage, such as rock carvings and rune stones, a great advantage of the device is that the measurement is contact-free. The in many cases fragile surfaces of these objects are not affected. Since the laser scanner records only heights, another advantage is that there is no interference from the colour of the rock. When used as a tool for reading runic inscriptions the best results have been achieved on surfaces having a fairly rough micro-relief. These are, for example, the inscription **braido** (Figure 4.3), the Skramle stone (Figure 4.6), and the Kuli stone (Figure 4.7). The reason might be that the eye is easily disturbed by such surfaces, while objective measurements by the laser scanner always pick out the true height values for every measured point. As the height values are stored in text format directly in the field on a portable computer, another important advantage of the equipment is that the data are available for further treatment and visualisations immediately after a scan is completed. One disadvantage is that some parts of the frame of the machinery are susceptible to rusting. However, this has only been a problem when measurements have taken place on the seashore, near water with high salinity.

The readings from a laser of the triangulation type are reliable and results are excellent when, for example, measuring inscriptions and rock carvings. There are usually no abrupt height differences on the surfaces and the number of unwanted reflections is negligible. The method is thus adequate for most types of inscriptions, while alternative techniques have to be sought when objects with a more pronounced relief are examined.

For documentation of more extensive inscriptions the measurement procedure has to be faster. It is possible to construct a new laser scanner, based on the same principles, to perform measurements at a speed five to ten times higher than the scanner used for the micro-mappings in this article. This will allow a surface of about 5 square metres a day to be surveyed with the same resolution and accuracy as at present. In many cases there is an urgent need for detailed documentation before the items weather away or decay by other means. Other improvements that can be made to the equipment are a reduction in weight and a further facilitation of the practical use of the machinery. The equipment has to be easily transported in the field and to different museums, preferably by only one person. The method used for describing the rune texts in this article can thus become one of the common techniques when reading and interpreting ancient inscriptions both in the field and in archives.

Note

Gunnar Berg constructed the electronics and Magnus Jansson wrote the software for the data acquisition and control of the motor boards of the laser scanner. Caroline Patjas corrected the English in the article.

References

FRIESEN, O. VON (1924), *Rö-stenen i Bohuslän och runorna i Norden under folkvandringstiden*, Uppsala Universitets årsskrift: Filosofi, språkvetenskap och historiska vetenskaper 4 (Uppsala): 1–165.

FRIESEN, O. VON (1933), *Runorna* (Stockholm): 1–264.

GUSTAVSON, H. (1996), 'Undersökning av runstenen vid Skramle, Gunnarskogs sn, Vä', unpublished report, Riksantikvarieämbetet: 1–5.

GUSTAVSON, H. (2003), 'Oklundainskriften 70 år efteråt', in W. Heizmann and A. van Nahl (eds.), *Runica, Germanica, Mediaevalia*, RGA-E 37 (Berlin and New York): 186–98.

HAGLAND, J. R. (1998), 'Innskrifta på Kulisteinen: Ei nylesing ved hjelp av Jan O. H. Swantessons mikrokarteringsteknologi', in A. Dybdahl and J. R. Hagland (eds.), *Innskrifter og datering/Dating Inscriptions*, Senter for middelalderstudier Skrifter nr. 8 (Trondheim): 129–39.

KITZLER, L. (2000), 'Surface structure analysis of runic inscriptions on rock: a method for distinguishing between individual carvers', *Rock Art Research* 17/2: 85–98.

KRAUSE, W. (1966), *Die Runeninschriften im älteren Futhark*, Abhandlungen der Akademie der Wissenschaften in Göttingen: Philologisch-Historische Klasse, 3rd ser., 65 (Göttingen): 1–328.

NORDÉN, A. (1931), 'Ett rättsdokument från en fornsvensk offerlund', *Fornvännen*: 330–51.

ODENSTEDT, B. (1997), 'The Skramle inscription: an important find with some notes on the older runic inscriptions', in M. Rydén, H. Kardela, J. Nordlander, and B. Odenstedt (eds.), *From Runes to Romance: A Festschrift for Gunnar Persson on his Sixtieth Birthday* (Umeå): 165–80.

SALBERGER, E. (1980), *Östgötska runstudier* (Göteborg): 1–42.

SWANTESSON, J. O. H. (1989), 'Weathering phenomena in a cool temperate climate', *GUNI Rapport* 28 (University of Göteborg, Department of Physical Geography): 1–193.

SWANTESSON, J. O. H. (1992a), 'A method for the study of the first steps in weathering', in G. Kuhnt and R. Zölitz-Möller (eds.), 'Beiträge zur Geoökologie', *Kieler Geographische Schriften* 85: 74–85.

SWANTESSON, J. O. H. (1992b), 'Recent microweathering phenomena in southern and central Sweden', *Permafrost and Periglacial Processes* 3: 275–92.

SWANTESSON, J. O. H. (1994), 'Micro-mapping as a tool for the study of weathered rock surfaces', in D. A. Robinson and R. B. G. Williams (eds.), *Rock Weathering and Landform Evolution* (Chichester): 209–22.

SWANTESSON, J. O. H. (1996), *Mikrokartering av hällristningar i södra och mellersta Sverige*, Högskolan i Karlstad, Arbetsrapport Naturvetenskap/Teknik 96/2 (Karlstad): 1–41.

WILLIAMS, R. B. G., SWANTESSON, J. O. H. and ROBINSON, D. A. (2000a), 'Measuring rates of surface downwearing and mapping micro-topography: the use of micro-erosion meters and laser scanners in rock weathering studies', in H. A. Viles (ed.), *Recent Advances in Field and Laboratory Studies of Rock Weathering*, Zeitschrift für Geomorphologie, Supplementbände, Band 120: 51–66.

WILLIAMS, R., ANDRADE, C., FOOTE, Y., FORNOS, J., LAGEAT, Y., MIOSSEC, A., MOSES, C., ROBINSON, D. and SWANTESSON, J. (2000b), 'European shore platform erosion dynamics (ESPED): monitoring the downwearing of shore platforms in the context of the management and protection of rocky coasts', *EUR 19359 — EurOCEAN 2000: the European Conference on Marine Science and Ocean Technology. Project synopses*, vol. 2, *Coastal Protection: Marine Technology* (Luxembourg): 499–503.

CHAPTER 5

Virtual Reality, Relative Accuracy: Modelling Architecture and Sculpture with VRML

Michael Greenhalgh

The web continues to grow as a vehicle for research and teaching, and the expectations of users continue to outstrip the ability of the technologies to deliver. 'Flat' two-dimensional images are easily stored, transmitted, and displayed, but lack the context inherent in the actual objects and their surroundings. How can the tools of the web be used to simulate the real world, and should we expect any software innovations usable by lecturers (rather than by teams of highly skilled computer specialists) in the immediate future?

These questions are of interest to those in many areas where varieties of distance learning are involved, from art historians and archaeologists to geologists and museum specialists who aim in some way to integrate objects into a context visitable today, or to reconstruct elements of some past state.[1] Although there is no substitute for viewing the actual objects in their intended location, this is frequently not possible — hence another reason for the development of 'computerised' contexts, echoing related attempts (for example with stereo photography)[2] in the analogue world.

My interest in the subject of this symposium is as an art historian who, since January 1994, has run a website (ArtServe) which now contains over 160,000 images of works of art and architecture for teaching, and who wishes to extend the supply of information by finding ways of providing students with a heightened sense of reality, either by providing context or by conjuring something more lifelike than the familiar flat images on the computer monitor, itself but a recent variation of the usual Renaissance 'window' analogy. Just as we can look out of a window, so we can look into a computer monitor, but the glass of the screen is a barrier which reinforces the flatness of what we see. Unlike Alice, we cannot go through this particular glass and interact with the invented world on the other side (although, with a wearable computer,[3] we can take advantage of multimedia[4] when we are actually on site — useful for art historians, archaeologists, and a host of training

activities).[5] Conversely, similar technologies can be used in invented contexts — for example where archaeologists execute a virtual dig.[6]

The web is the best available carrier and broadcaster of visual information and, in spite of the difficulties rehearsed below, the best bet we have for developing alternative worlds, both actively, in the expository set-up of the lecture hall (via a networked computer feeding a video projector), and passively, by having students of the discipline visit appropriate websites. The web is preferable to stand-alone programs for the construction of graphical presentations for reasons of access and cost (two of the gods of much university teaching), but also because web technologies impose the simplification of a lingua franca (albeit sometimes rough and ready) on the design and use of projects.[7] If a presentation works on my set-up, it should work on everyone else's. Accessibility allows the discipline and its students to reap the financial and pedagogic benefits available, according to some, from the development of distance learning (although others believe the technology too expensive, and actually distracting).[8] The funding argument is that, given the high cost of establishing, maintaining, developing, and using traditional slide-based image collections (which have disadvantages for students compared with the web),[9] flexible forms of digital imaging might provide the only means of maintaining and improving pedagogical procedures in an era of diminishing funding. It is ironical that, amid the wealth and technological riches of the First World, so many institutions should be contemplating launching into web-based teaching because it is perceived (possibly wrongly on both counts) as a money-generating milk-cow or a way of mitigating declining

---

[1] See Barceló et al. (2000) for an overview in archaeology; Humanities VR links at http://www.theatron.org/links.html; archaeology links at http://www.gla.ac.uk/archaeology/projects/SSEMS_web/resource.html.

[2] Cahen (1989); Peretz (1989).

[3] http://www.media.mit.edu/wearables/.

[4] Broad introductions are in Brice (1997); Furht (1999); Kurbel and Twardoch (2000).

[5] VVECC (2000): David Johnston and Adrian Clark 'are developing an augmented reality tour guide application for such sites which uses a wearable computer with GPS positioning capability, driving a head-mounted display with orientation sensors. This system is able to position the wearer at the correct place in a 3D model of the site as it is likely to have appeared in the past, greatly improving the experience'; cf. the great interest in VRML evinced in the University of York's refereed online journal, *Internet Archaeology*, at http://intarch.ac.uk/; overview in Dong and Gibon (1998).

[6] Papaioannou et al. (2001).

[7] See Debevec (2001) for a survey of projects 1991–2000.

[8] Young (2000).

[9] Accessibility, ease of use, freedom of navigation, and image quality, say Ludlow and Platin (2000), after a comparison between web and slide/tape set-ups for student self-paced learning.

staff numbers, rather than as a valuable opportunity to improve and enrich learning methodologies.[10] Needless to say, research in this area of computing is expensive; only a few of the results are prima facie useful to non-specialists; and expectation runs ahead of delivery, as it always has done in computing.

The establishment of context requires large amounts of information on setting, structures, and objects to be served across the web. Driven by appropriate computer software, it includes the concept of immersion, and holds out the hope of manipulation, the acronym being IVR, 'immersive virtual reality'. That is, so much information is provided that the user should have the sensation of standing inside the ensemble — perhaps an archaeological site,[11] a museum,[12] or a church[13] — and be promised the ability to move through the spaces of this immersive world the better to appreciate immovable buildings from any angle and, concurrently, to have the computer manipulate objects as if they were in the hand. In prospect, users would be able to interact and collaborate within such virtual environments[14] — a useful tool in teaching. Several institutions around the world have tested the concept of immersion by building cubes fed by back-projection (inspired by the first device at the University of Illinois), in which the user stands and manipulates the computer feeds and hence the environment[15] (although this cannot of course be web-based — another indication of the barrier erected by the glass of the computer monitor). A recent survey acknowledges that the present state of the art is primitive and very expensive, but argues that 'evolving hardware and software technology may [not will] enable IVR to become as ubiquitous as 3D graphics workstations — once exotic and very expensive — are today'.[16] In the more optimistic visions, people can actually meet within virtual museums.[17] In the heritage dimension, it may well be that virtual reality is as close as most of us will come to endangered sites, and this can apply to dive-sites, which can also be created virtually.[18] Again, VR can perhaps make the monuments more attractive,[19] and may form one of the methodologies in the ambitious digitisation of European heritage,[20] especially

as some researchers, noting the deficiencies of current techniques, examine laser scanning.[21]

For the lecturer, the main interest of VR techniques is in the re-creation of three-dimensional works, which can never be adequately treated on the flat pages of a book or on a lecture screen. Architecture is almost impossible to convey successfully in the lecture theatre, because even a succession of individual shots, and even when these are keyed in some way to some kind of floor plan for clarity (see below), can never adequately conjure up the effects of space and volume with which the architect works. But context can also enhance the student's understanding of flat works such as paintings and drawings, because many of these can be understood only in a setting — whether the manufactured setting of an art gallery, or the original or reconstructed context of a palace — where the target takes its place amid other works of art, which can help explicate it.

This paper will assess, with the scepticism of experience, the current possibilities for the attainment of realistic context over the web by trying to match the basic requirements of art history scholarship and teaching against what is currently offered and what can be confidently expected in the near future. To do so, it surveys some ongoing research in the field from the perspective of an outsider and a user. It is divided into two sections. In the first, dealing with VRML (virtual reality modelling language), it describes one project completed in collaboration with the Supercomputer Group at the Australian National University to model in VRML the Buddhist stupa at Borobudur (built about AD 800), and then refers to a second project dealing with Piazza del Popolo, Rome, and the reasons why it was decided not to implement this second project in VRML. The restrictions of current software for the 'automatic' generation of VRML models are discussed in lay terms, and the conclusion is that many hurdles remain to be overcome before virtual reality can (without very great effort and/or sophisticated and bulky equipment) reflect with accuracy the detail of the real world. Hence it questions the power and utility of this particular technology, and its potential in teaching art history (or any discipline where the web is to be used to model three dimensions), at least within any sane funding models where hard choices have to be made. The second section looks at other ways in which the ordinary lecturer (rather than the computer specialist) may employ various simple technologies to conjure context, and with more flexibility, detail, and accuracy than VRML can yet (or probably ever) be made to achieve.[22]

VRML and the Construction of Ancient Worlds

As computers increase in power, user expectations nevertheless outrun reality: viewing still images begins to lose its attraction as the possibilities of more sophisticated media such as video and constructed

---

[10] Hall et al. (1999).

[11] Cf. the Virtual Worlds in Archaeology Initiative: http://www.learningsites.com/VWinAI/VWAI_participants.htm.

[12] See Marty (2000) for planning virtual museums over the web.

[13] Frischer et al. (2000).

[14] Greenhalgh (1999).

[15] E.g. the VR-Cube in the Centre for Parallel Computers at KTH in Sweden, which has six walls, 3 × 3 × 2.5 m high: http://www.pdc.kth.se/projects/vr-cube/.

[16] Van Dam et al. (2000), 26.

[17] Paolini et al. (2000): 'We briefly illustrate, mainly through pictures, a prototype application, which is a virtual tour within a portion of the Museum of Science and Technology of Milan.'

[18] See Kenderdine (1998) on the Western Australian Maritime Museum.

[19] Berndt and Teixera (2000): they instance Santa Clara-a-Velha, Coimbra, at http://www.ccg.uc.pt/projectos/virtual/santaclara.html.

[20] Cf. http://www.hatii.arts.gla.ac.uk/research.htm.

[21] E.g. Addison and Gaiani (2000), for their work on Pomposa.

[22] *IEEE Computer Graphics* 19/2 (1999) is dedicated to VRML.

'worlds' appear to come within our reach.[23] However, we should recognise that there is a gulf between those applications where VR must be used because there is no effective substitute (such as in surgery),[24] or where the set-up cost is much less than that for any alternative available technique (such as vehicle manufacturing, architecture,[25] or the training of pilots or astronauts), and cash-strapped disciplines such as art history, where the awkward requirements of detail and accuracy in a cathedral interior are more exacting than for a river estuary at dusk, an airport layout, or the shape of a new motor car (and the more detail, the larger the files — although this problem will lessen as networks and machines get faster). Dealing with buildings in the real world, with all the vagaries of asymmetry, weather, lighting, intricate detail, and convoluted shapes, is a difficult proposition. Can art history be unique in a stipulation of quality in VR? From the web, it seems clear that archaeological reconstructions can offer abbreviated results that are acceptable to an audience already used to the sparse blocked-out reconstructions traditionally provided by archaeological draughtsmen because so often the information necessary to include detail is lacking.[26] To move from sketch reconstructions to detailed ones requires a lot of work (as in the admirable Pompey Project, Chapter 7),[27] but detail is not necessarily the same thing as accuracy. To admit the deficiencies of VR is one thing, but we should deplore the kind of cop-out which suggests VR as a parallel world to the real one, which can have its own rules, and therefore conveniently does not have to approach reality.[28] It is a denial of the respect for detail and accuracy that scholarship demands in any discipline, whether it appears on the printed page or on the web. In this respect, perhaps VR shares the common mistrust people have for still digital imagery: the computer monkeys about with the truth, so no digital image is to be trusted. A more accurate term such as simulation[29] is therefore preferable to describe the *proxime accessit* that this technology delivers — although one report maintains that virtual reality offers a qualitatively better way of teaching about painting than simply showing a 2D image.[30]

The features available in VRML highlight the apparent attractiveness of the technology,[31] beyond what an actual site visit could supply, offering levels of interaction very useful in teaching and learning. These include hyperlinks to *comparanda* from other contexts, to high-quality images, audio or video clips, textfiles, and so on, or to searchable databases of apposite materials, viewed in the margin or in a separate window. The scaling and/or quality adjustment of the whole model to suit various speeds of network or computer is also attractive.[32] Unfortunately, these advantages are counterbalanced by inherent drawbacks which ensure that VRML projects share a family resemblance. They are poor on details (programming difficulties); they are fuzzy (the size of the imagemaps has been cut down to ensure smoothness of transmission); they repeat details (to save programming time); and they don't look real (their colouring is often washed-out and does not duplicate the richness of the real world).[33] The optimistic answer is that this is because of a lack of tools to quickly, inexpensively, and accurately model reality[34] — although some projects, such as one involving Notre Dame, Paris, do indeed promise accuracy,[35] and the UCLA Cultural VR Lab states its mission is to create highly accurate 3D computer models of culturally significant sites around the world.[36] The pessimistic answer is that VRML cannot deliver accuracy and detail for large ensembles in any acceptable time frame and cost.

VRML can be pictured as a two-stage process in which a representation of the outside of the structure — the outer skin, as it were, often derived from photographs of the actual structure — is hung on a geometrical model. Hence there are two distinct tasks to be performed: namely the construction of accurate geometry, and its clothing with the detail of the real world — with what we see when we look at the real structure. The model may be constructed by entering detailed measurements into a CAD (computer-assisted design) package (although this is not necessarily snag-free).[37] Since each visible surface is a polygon, procedures are available to clothe each one in real-life textures, and the product can then be exported in VRML format.

Manipulating a CAD package to produce an accurate and detailed model is a highly skilled task, and various photogrammetry programs offer what might appear to be an easier alternative which deals with

[23] See Durlach and Mavor (1995), which has a good early bibliography; Singh (1994) for early consideration of some basic technologies.

[24] Mangan (2000), at Penn State.

[25] See Porter and Neale (2000) for an overview of ways of building architectural models.

[26] Novitski (1998) largely features Antonieta Rivera's reconstruction of Teocalli, the ceremonial precinct of the Aztec capital of Tenochtitlan; cf. Novitski (1999) and Sagalassos: http://www.esat.kuleuven.ac.be/sagalassos/.

[27] Summary at http://www.nyu.edu/its/humanities/ach_allc2001/papers/beacham/: 'the 3-D models acquire a "persuasiveness" which can easily render invisible to the viewer crucial distinctions between known fact, scholarly deduction, and creative (albeit educated) guesswork'.

[28] Heim (1998), 48: 'Realism in virtual reality refers not to photorealistic illusions or representations. Reality also means a pragmatic functioning in which work and play fashion new kinds of entities. VR transubstantiates but does not imitate life.'

[29] Müller and Quien (1999).

[30] Antonietti and Cantoia (2000).

[31] See Chen (1999), 175–211, for a *tour d'horizon* of virtual environments.

[32] For example, as Bologna does with its NuME models: cf. http://www.cineca.it/nume/.

[33] An overview of creating virtual environments at the VVECC (Visualization and Virtual Environments Community Club) Design of Virtual Environments meeting, 30 January 2001, is at http://www-ais.itd.clrc.ac.uk/VVECC/proceed/vedesign/.

[34] Addison and Gaiani (2000): the paper also gives a *tour d'horizon* of projects.

[35] DeLeon (1999): 'The results derived from this accumulation of data will be precise and as scientifically accurate as possible within the means of our employed technologies.'

[36] http://www.cvrlab.org/humnet/; background to the lab at http://www.cvrlab.org/media_articles/labusiness.gif.

[37] Brooks (1999); cf. p. 19 for problems with acquisition and translation, data cleaning, and accurate rendition of polygons.

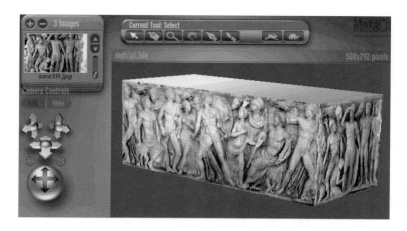

Figure 5.1 VRML (Canoma): a Louvre sarcophagus mapped onto appropriate geometry.

'structure' and 'clothing' at the same time. With such software we begin with photographs of the structure from all angles, and then 'tell' the software where the various surfaces, angles, and profiles are to be found on each photograph. The software then constructs the model from the co-ordinates it has learned. The goal is obviously the automatic extraction of as much information by software as possible, with as little work as possible to be done by the human being, and several packages proclaim increasingly high levels of such automation, including automating the task after first having the user set down the basic parameters — 'model refinement', as Anthony Dick has called it. It is even rumoured that three-dimensional information should be capable of extrapolation from stereoscopic photographs, but I have never seen this demonstrated. At present 'totally automatic' model extraction needs large numbers of photographs, and a lot of computing power. Could somebody point me at a model of a work of architecture which uses such a technique? If the evidence exists in the form of a model thus constructed, it is very well hidden.

Superficially at least, this 'working backwards' from photographs of the real world to a geometric construction that can be imported into VRML should be much more lifelike than going the other way, via CAD. This is because the photographs from which the geometric information is extracted are already clothed in the correct textures and hues, and set in the true environment — so that all the user has to do is to clothe the constructed 3D version with the real-life textures, usually using images which start out uncalibrated, and taken from arbitrary viewpoints. Indeed, for simple structures, the technique works well, even if sometimes the structures must be dirtied to make them look real rather than artificial,[38] and are supplied with 'environmental features'.[39] We may bless the design philosophy of Le Corbusier or Mies van der Rohe as we construct versions of the *machine à habiter* or the conventional tower block remarkably quickly. All the angles are right angles; windows are square; structures are cubic; details are the same repeated from window to window or from floor to floor.

But this would be a tedious procedure indeed for complicated surfaces such as baroque architecture (of, for example, Bernini or Gaudì or Vanbrugh), or for any kind of sculpture. Figure 5.1 shows how such a photogrammetry package deals with a sarcophagus in the Louvre, forming only a vague approximation of the contours. Such software is unable to handle non-cubic shapes easily, unless the user tediously plots every smallest variation.

Unfortunately, then, there are insoluble snags to starting from photographs that are not easy to overcome, and which one doubts software could ever solve automatically.[40] They fall into several categories.

1   The software needs the human being: we may think of the software as blind to the photograph, and the human being as the guide-dog that simply offers the software a collection of points similar to map references in two dimensions identified and located on at least two photographs (three or four give better precision) taken from different angles, which the software can extrapolate into three dimensions. The more information the human being feeds in, the more accurate the finished model can be. Unfortunately, we are a long way from any software 'recognising' corners or even verticals, let alone measurements of distance — so sitting back for five years until the software can do the lot does not seem to be an option.

2   Orientation: photomodelling programs are very choosy about the relative angles of photographs, and cannot extrapolate 3D information if the views are incomplete or the camera positions too close together; yet, in many real-life situations, good angles are simply not attainable.

3   Fields of view: life would be simple were each target building on a podium, with a large field of view all round, and convenient hills from which to take top-down shots (which is indeed the way in which some VRML projects tend to be presented!). But most buildings are obscured — occlusion — by anything from street furniture to other buildings or trees. Photography from all angles is rarely possible, and

---

[38] Warniers (1998a).
[39] Jin and Wen (2001).

[40] See the survey of image-based modelling programs in Warniers (1998b).

the golden rule of photomodelling programs — to take shots at 45° looking towards the corners — is sometimes very difficult to follow.

Two consequences inevitably flow from such snags. First, the more accurate the detail, the more work is required: using sections of the photograph to 'clothe' the model is progressively more difficult the more detailed and twisty-curvy the object. To forget the detail and go for the 'grand picture' results in cardboard cut-outs, just like a piece of stage scenery — a 'flat'. Second, repetition is required in order to save work: the modelling of difficult objects (columns, capitals, individual cofferings in ceilings) is often done once, and the same item repeated down the whole façade, or across the whole ceiling. This might look superficially convincing, but is of course far from accurate (unless for a model of something like the Fushimi Inari Shrine at Kyoto, where there are indeed thousands of scarcely varying *torii* gates).[41] Even with elaborate projects, such as the Urban Simulation Team's modelling of Los Angeles,[42] it is easy to spot where repetition has been used, which will certainly restrict the uses to which the models can be put.

With such problems and commonly accepted 'solutions', what level of stage scenery or fakery is acceptable or noticeable? Of course skilled programmers can work wonders, but any comparison of such a VRML model with the real thing, or even an ordinary photograph, exposes the simplification. We no longer believe in the 'realism' of photographs, but in VR any rigorous definition of realism has little part to play — and this in spite of the dream that a few mouse-clicks will build a model almost automatically.

But can software really perform sophisticated functions at the click of a mouse?[43] Programs can often construct very acceptable panoramas and stereo pairs, but detailed VRML constructions are more difficult, with no software in sight which can make them much easier. Improvements in appropriate software (and there have been many) have been in automating the necessary processes (not changing them) to cut down the time required to produce a quality product, but not a detailed and accurate model. There is an inevitable disparity between production software and the laboratory; thus in 1999 it took several human-years 'to make a model of an existing kitchen that aims at quarter-inch accuracy',[44]

while one current team presents a technique to obtain texture-mapped models of real scenes with a high degree of automation using a video camera and an overhead projector.[45]

Proof of the meagre help offered to ordinary users by automation is demonstrated by the paucity of sophisticated and scholarly VRML presentations visible on the web or on CD-ROM. Large VRML programs cost money, for extensive teams are required. The model of the Cathedral of Santiago de Compostela done for the Santiago European City of Culture 2000 by the University of Santiago (Laboratorio de Sistemas del Instituto de Investigaciones Tecnológicas)[46] took over 100,000 hours of work, and employed over twenty people (including two project leaders, three programmers, three on 3D design, two on digital photography, and one on stereography). The result offers superb programming devices, such as changing reflections on a sliding glass door, apparently accurate capitals and sculpture inside the cathedral (in fact they are copied and multiplied), and an all-too-familiar storm with thunder and lightning in the square outside the cathedral. But it does not offer accuracy or detail sufficient to use for extended teaching. Granted that large VRML projects sometimes involve political aspects (using computers is 'modern', so universities like to brag about them; visualisation labs need data and collaborate to prove their relevance; few results of computation look so pretty; culture can compete with science for funding), nevertheless to see the paperwork for large projects would be intriguing — almost as much as asking what the targets were, and whether and how they might have been achieved more cheaply. Could, for example, a richer and more convincing range of documentation for both art historians and the general public have been achieved by using some or all of the alternatives to VRML that I suggest later in this paper?

Automating Capture: The Problem of Size

The strictures above relate to life-size buildings and sites, not to small objects where (as we might expect) a true automation, using cameras and lasers linked to a computer, allows the measurement process and hence the construction of a VRML model to be automated. In such set-ups, the camera is fixed, and the object to be modelled rotates on a turntable — which proves that VRML can indeed provide both accuracy and detail: it is only the human intervention necessary for larger models which is fallible and exhausting.[47] The procedure offers a useful way of providing accurate models for online shopping, and also for smallish and relatively simple archaeological

---

41 Plante et al. (1999).

42 http://www.ust.ucla.edu/ustweb/projects.html — the same team has made reconstructions of the Forum of Trajan, Rome; cf. Delaney (2000).

43 'Soon, people will no longer need to go beyond the realms of their own homes to experience or study intricate heritage structures': cf. Haval (2000), and the project treating the palatial complex of Fatehpur Sikri, at http://rohini.ncst.ernet.in/fatehpur; or cf. a 1999 review of a UCLA project as 'Rome virtually reborn at click of mouse': http://www.humnet.ucla.edu/humnet/classics/faculty/frischer/Reuters.html.

44 *IEEE Computer Graphics* 19/6 (1999) is dedicated to virtual reality; quote from F. P. Brooks, 'What's real about virtual reality?', at pp. 16–27: see p. 19.

45 Czernuszenko et al. (1999); we may perhaps look forward to samples being made available over the web.

46 http://www.cidadedacultura.es/.

47 Cf. the work at Fraunhofer Institut Graphische Datenverarbeitung: http://www.igd.fhg.de/igd-a7/. Other 3D techniques also work in miniature. *IEEE Computer Graphics* 20/2 (2000) is dedicated to perceiving 3D shape: see R. Schubert (2000).

artefacts,[48] but is of little use to art historians because of the impracticality of treating more than a few works, such as Michelangelo's 17 feet high *David*.[49] What is more, most art historians work alone, on a low budget, so their equipment (cf. the four tons of hardware shipped for the *David*)[50] generally needs to be portable and unobtrusive. Conceivably, a program for dealing with appropriately sized art objects — Renaissance statuettes, porcelain figure groups, arms and armour, or Greek terracottas — could offer interesting results, and would not be expensive. Thus the smallest objects might be covered by something like the Fastscan Handheld Laser Scanner, which generates a profile from overlapping sweeps of the wand.[51] Naturally such a system is restricted to objects the human arm can wave over, although larger devices have been used in conjunction with a turntable.

If such devices can offer VRML models of small, discrete objects (with probably a big future in medicine and surgery),[52] a different approach is needed for dealing with room interiors. To obtain a suitable variety of images from the requisite angles, an automated trolley bristling with cameras like a helicopter-gunship has been developed, with an on-board computer to store the large amounts of data.[53] Processing that data is currently non-trivial, and such a set-up is naturally expensive. Inspired perhaps by the automata used in dangerous locations, such machines no doubt have a future in industry and perhaps with the military; but for the rest of us, such devices[54] are impracticable outside the laboratory or specially equipped rooms.

Borobudur in VRML

A well-known problem with computers is that we tend to accept what their software and capabilities can offer us. This is natural, because if we wish to use a computer we have no real alternative unless we have money, programmers, and plenty of time. By so doing, however, we tend to compromise on our requirements — which is dangerous. In other words, technology determines use, whereas the basic requirements of any discipline specifying the required output should determine the technology. This inversion of roles is seen in many VRML projects, including Borobudur.

The stupa of Borobudur[55] was chosen for VRML attention because it presented several features which made it easier to do than — say — St Peter's or Blenheim. It is a solid block (there are no interiors), and it has a regular, more-or-less cubic structure, so that each side echoes the other three, in a series of pyramid-like steps giving access to galleries and eventually to the summit. A significant feature of the structure is the more than 2000 relief panels, all available in out-of-copyright photographs of excellent quality — and all in greyish monochrome, which is the natural and unattractive tint of the andesite from which they are made (the reliefs were probably once coloured over a stucco ground, but all traces have disappeared). Plentiful photographs of the whole stupa, and some aerial ones (taken by the Dutch last century), offered sufficient information for conversion to VRML.

In the finished VRML model, the reliefs look good — almost three-dimensional — because the original photographs were well lit, and it was a relatively simple task to hang them as flat pictures in a computer-constructed art-gallery. Any navigation around the four storeys of galleries of the model (helped by a gauze overlay showing current position: cf. Figure 5.2) which concentrates on the reliefs creates a good impression. Look up, however, or zoom into the architectural framework and the three-dimensional Buddha figures it holds, and the illusion collapses. Full detailing of the architecture was omitted because we simply lacked the necessary time and money, and because we doubted whether a realistic effect could be achieved. Instead, as already explained, one section has been modelled, and then copied — so the result is no use to anyone wishing to study the monument's subtleties, or to be assured of the accuracy of what they see in the model, because on the actual stupa each Buddha figure and each architectural feature is different from its fellows. In effect, then, and notwithstanding the 'cardboard cutouts', the 3D sculpture — a fair proportion of the interest of the monument — has been ignored. And since the monument has over 300 sculptures in the round, this is a serious deficiency.[56]

Extra programming was also required to cater for the quantities of data required to 'feed' the galleries with their reliefs (and all from a CD-ROM, or across a network). The monument was divided into over twenty sections, and three different resolutions of each section were generated — so in effect there are about seventy different VRML segments to cater for network and machine speeds. Figure 5.3 shows that the user may easily jump to another segment by clicking on the elements of the gauze overlay. As networks and machines get faster, so higher-resolution images can be used to

---

[48] E.g. at SEMSS (Scotland's Early Medieval Sculptured Stones): http://www.gla.ac.uk/archaeology/projects/SSEMS_web/.

[49] Levoy et al. (2000); cf. the survey in Shulman (1998).

[50] Some statistics are at http://graphics.stanford.edu/projects/mich/.

[51] A brief review is at http://cadence.advanstar.com/2000/0200/tr0200_fastscan.html; cf. Schubert (2000) for a stereo scanner; http://www.computersculpture.com/Computer_resources.html lists several similar devices.

[52] Cf. the Virtual Pelvis Museum at Manchester: http://www.hpv.informatics.bangor.ac.uk/Sim/Pelvis/; also the VVECC December 2000 meeting, Virtual Environments in Medicine and Psychiatry, at http://www-ais.itd.clrc.ac.uk/VVECC/proceed/medical/.

[53] For example, Professor El-Hakim's mobile mapping system, at http://www.vit.iit.nrc.ca/elhakim/.

[54] To which we may add holography: Beiser (1988); Kransteuber (2000): I have not seen this item.

[55] The Borobudur presentation, including a database, various files of information, and the VRML model, is available at http://rubens.anu.edu.au/htdocs/bycountry/indonesia/borobudur/.

[56] *IEEE Computer Graphics* 17/2 (1997) is dedicated to 3D and multimedia on the information superhighway.

Figure 5.2 Borobudur VRML project: a gallery with gauze location overlay.

Figure 5.3 Borobudur VRML project: a gallery with a gauze overlay for selecting a different tour.

clothe the model's armature. We ran out of time, money, and resources, but the Borobudur Project demonstrates how VRML can be integrated into a project which also offers other ways to study the material. Thus, all statues, reliefs, and architectural features are available as a series of 2D photographs. Were we to start the model again, we might well consider using VRML only as a block model, employing other techniques (see below) to represent detail and accuracy. Hence the VRML model might be used as a finding aid or springboard to various other generally simpler technologies.

An index of the awkwardness of even the best photogrammetry software is that the programmer of Borobudur, Dr Ajay Limaye, chose to construct the model by hand from a series of prepared primitives, rather than to extrapolate the model from a series of photographs. This was in part because the photographs we used were not taken with any kind of photogrammetry in mind, and would therefore have had to be carefully matched before they could be used, but also because attacking the VRML directly provided greater leeway of action for the programmer. To use photogrammetry would have required a visit to the site (Java faced political unrest when such a visit was feasible). Indeed, to capture the wide-angle shots necessary for such work is only now becoming possible with prosumer digital equipment, as only recently have affordable digital cameras appeared which will take interchangeable lenses (the widest angle on a fixed-lens zoom is currently the Minolta Dimage 7, with a 28–200 mm lens).

The Barrier of the Computer Screen

Part of the problem with VR methodologies is that they are not helped by the flat nature of the computer monitor which, I have suggested, behaves rather like the Renaissance window in that in order to view the 'world' we must look through the glass, which acts more as a barrier than as an aid to immersion. This was easily demonstrated by splitting the Borobudur signal into left and right sections and rear-projecting them (also split into a false stereo in order to add apparent depth) onto the Australian National University's more restricted version of the cube, called the wedge—namely 4 m long screens placed at right angles. The viewers, equipped with shutter glasses and a navigation device, stood within the right-angled V of the screens, and controlled the movement (including zooming) of the model around them. Standing within the model, the impact of the architecture is enhanced both by being there, and also by the larger size of the projected artworks. The high-quality digital images now possible can only appear to good effect on wall-sized displays,[57] such as UCLA's Visualization Portal.[58] Very soon now (if not sooner), TFT-like screens may be sold by the yard like wallpaper and pasted on walls, so the ordinary monitor becomes obsolete.[59]

[57] Funkhauser and Li (2000). Several papers on similar themes are in the same issue of *IEEE Computer Graphics and Applications*.
[58] http://www.ats.ucla.edu/portal/.
[59] For a survey see Van Dam et al. (2000), 37–8.

Indeed, it is difficult to see how IVR can otherwise develop.

My current project, on the monuments in the Piazza del Popolo, Rome (an area which boasts three churches, an obelisk, two fountains, a city gate, and the city walls), will use VRML, if at all, only for one of the entrance points into the project, in the form of a block model. More effort will be put into the generation of panoramas (some of them linked to floor plans) and imagemaps, because excellent results with such techniques can be produced relatively quickly and without learning a complicated software package. Equally important, detail and accuracy can be preserved, albeit at the expense of three dimensions.

In any case, it will be apparent from the above that I do not consider VRML able by its nature to provide accurate and detailed models of the world as it exists (archaeological reconstructions, some might argue, are a different matter). As an eye-catching sales-pitch for hardware manufacturers and leading-edge museums and cities they serve a purpose, but the effect is sometimes more Disneyland than reality. A scan of the web does not indicate such models yet in everyday use as learning aids or as the subject of scholarly study, surely because of the deficiencies of the technology. Given the amount of attention paid to VRML, if there were a software solution to this problem, it would have been apparent by now; but since the bottleneck appears to be neither computing power nor speed, and since research concentrates on objects that fit on a turntable, we should for the time being look elsewhere.

Alternatives to VRML for Providing Context

Is VRML the only technology available that can provide both context and more information than a flat two-dimensional image? No, because there are several ways a computer can display more of the world than is visible in one photograph, all of which are constructed by software in the computer, and most of which our predecessors in earlier centuries were keen to implement in some fashion, because they likewise recognised the restricting nature of two dimensions whether in paint or photography. After all, illusionism in art goes back at least to the Greeks. Viewed against such long-standing preoccupations, VRML itself, which might seem totally new, may be thought of as the computer equivalent of stage scenery, while CAD models are versions of the real thing, built in wood.

Panoramas

Large-scale panoramas were popular in the Renaissance, a veritable craze in the nineteenth century,[60] and are the subject of ongoing research in computing.[61] By using the computer to stitch overlapping photographs into a panorama, perhaps even in 360°, we can provide an image larger than the monitor/web browser display, and around which the user can scroll at will — a cheap alternative to large screens.[62] The panorama need not be a confection of separate adjacent shots: given the pixellation of which consumer-level digital cameras are now capable (from 3 to 4 megapixels, sometimes interpolated up to 6 or 7 megapixels), it could just be a very large digital image of which only a section appears in the browser window, and around which the user scrolls using cursor keys or mouse (once again, because of the limitations of the computer screen). If most panoramas are of stitched sequences (with a tripod necessary for indoor work), experiments with a suitably profiled, polished aluminium cone can give a 360° panorama in one shot, which can be 'flattened out' with software, but with two restrictions: (a) the very centre of the panorama — i.e. the apex of the cone — does not register, because this is where the camera is reflected, and so must be excised from the finished product; and (b) it is difficult to get very high-quality images, since every imperfection on the cone shows up on the finished product. Full 360° panoramas are also possible with fisheye lenses and special software. Even greater realism can be conjured by combining stereo colour anaglyphs with panoramas;[63] another technique, called 'omnistereo', involves special equipment and a panorama for either eye;[64] another, three cameras;[65] and a fourth derives stereo panoramas from only one camera[66] — the variety suggesting that such technologies are not yet ready for the ordinary art historian.

Advantages: the wider the view, the more information; panoramas can also be vertical (which is very useful for architecture: research work has derived panoramas horizontal and vertical from digital video),[67] allowing the viewer to see more detail than would actually be available on site without binoculars.

Disadvantages: a precisely registered sequence of photographs is required; the 360° version is tricky to perfect, hence the development of robotic techniques to make them without human intervention.[68]

Zoomable Panoramas

These provide better detail than ordinary panoramas, up to the level of the resolution of the image: they are the equivalent of the 'life size' panoramas of the nineteenth century (Figure 5.4), or indeed of stage sets, the details of which would be 'zoomed' with binoculars. We can treat any large images to zooming, using viewers

---

[60] Rice (1993), reviewing the Bonn exhibition of panoramas; cf. Oettermann (1997) and Comment (2000).
[61] Shum and Szeliski (2000).
[62] See   http://rubens.anu.edu.au/popolo/leadsto/panoramas/ for a selection of pans of Rome.
[63] Wei et al. (1999).
[64] Peleg et al. (2001).
[65] Yamada et al. (2000).
[66] Peleg and Ben-Ezra (1999).
[67] Nakamae et al. (1999): cf. figure 9.
[68] Mei and Wing (1999).

Figure 5.4 Nineteenth-century panorama: the Colosseum, begun 1824 (between Albany Street and Cambridge Terrace, London): seen here in a lithograph by Ackermann (1829).

such as PixScreen,[69] which allow the user to get very close before the image begins to break up. What about quality and web delays? Such panoramas can be packaged for download by the user, as self-executing files,[70] and they can be made interactive, as has been done with archaeological exhibits.[71]

Advantages: suitable machine/network combinations can handle very large images.

Disadvantages: the magnification is limited to the size of image the particular computer/network combination can handle without choking — so speed and ease of manipulation currently spell small, low-quality images.

Linked Panoramas

Linking by hotspotting makes ordinary or zoomed panoramas but one of a series of environments, with

movement between them by a mouse-click, as if moving around the rooms in a house: 360° panoramas are in use for library tours, which allow navigating, viewing, reading, hearing, and remote access,[72] and similar resources go into the making of campus tours.[73] Panoramas can also be linked with maps or plans within the same browser window, as in Figure 5.5, where Heemskerck's view of Rome is (anachronistically) linked to Nolli's map;[74] as the panorama is panned or zoomed with the mouse, so the light beam on the map moves around, its field of view reflecting the level of zoom.[75] Alternatively, the user may manipulate the beam directly, causing the panorama to swing around.

Advantages: the user is provided with the best available movement and all-round vision through a succession of linked spaces; added hotspots increase flexibility and the amounts of information available; maps/plans increase information for context.

---

[69] The program works either standalone (and can be bundled with the image into a .exe file), or as a Java applet on a web page: http://www.pixaround.com/.

[70] Examples from the Getty Center and the Palatine Chapel are at http://rubens.anu.edu.au/new/panoramas/usa/california/losangeles/getty/pix3/getty3.exe and http://rubens.anu.edu.au/new/panoramas/sicily/palermo/capella_palatina/pix2/palatina2.exe.

[71] Kriszat (1999): the author believes such a set-up avoids the 'lost in space' feeling.

[72] Xiao (2000).

[73] e.g. Mohler (2000).

[74] A variety of panoramas of Rome, some map-linked, are available at http://rubens.anu.edu.au/new/panoramas/italy.new/rome/.

[75] A Java applet, called *PMVR*, at http://www.duckware.com/pmvr.html.

Figure 5.5 Web imagemap: Heemskerck's panorama of Rome, linked by a Java applet to Nolli's map.

Disadvantages: the linked panorama is intricate if not difficult to set up, with a dedicated photographic campaign needed to collect the source images, and careful planning so as not to confuse the user.

Clickable Imagemaps

An easy alternative to zoomable panoramas is to construct an imagemap to make into a web page: this image is 'mapped' so that, when the mouse hovers over a particular reference area, related information is displayed in an adjacent window (as in Figure 5.6, where the details of the Chigi Chapel may be studied from its 'map' in the left window).[76] This could be an enlarged detail, a text file, or even a sound file. Further flexibility can be added by building in the code to accept a mouse-click as well, so that each hot area of the imagemap can accept two actions, rather than just the one. As may be seen from Figure 5.7, the technique works very well with sculpture[77] — an area where VRML is usually not an option. Imagemaps may also be combined with panoramas, for example as a way of offering a 'wall plan' of a building, with high-quality details, as in the wallpaper example (Figure 5.8) from Christ Church Cathedral, Oxford.[78]

Advantages: simple to set up, needing only a program to extrapolate the co-ordinates of the hotspotted targets. Conventionally, the imagemap can be small, to make sure it is always visible in its frame, but the images to which the hotspots point can be of very high quality. The illustrative images can be of any dimensions, focal length, and so on (that is, the rules required for constructing a panorama do not apply). Use is self-explanatory, and the imagemap is extensible with both mouse-hover and mouse-click to generate actions.

Disadvantages: the imagemap can work using two browser windows, but only on a monitor of sufficient size; normal set-up is with frames within one browser, which can seem a little claustrophobic on a small monitor.

Stereo Pairs

Stereo became popular with photographic experiments to simulate depth. Such images can look good on a computer, since they can be of any dimensions (hence of high quality); stereo effects can, with glasses, enhance the impression of depth. In an echo of cinema techniques of the 1950s, stereo movies of cultural heritage items have also been developed.[79] Since stereo contains depth information, enthusiasts make the link with photogrammetry and the complete automation of the mark-up process: '3D photography promises to make 3D modelling as easy as point and shoot', suggests one researcher[80] — yet solid evidence that this can happen is lacking, unless using elaborate kit to deal with smallish objects.[81]

Advantages: works well when targeted on one object in space. Software is available to blend and adjust the originals into one stereo image.

Disadvantages: red/blue glasses are required for viewing (but these are cheap); stereo bleaches colours, so does not work well with intricately coloured sculpture; it requires a precisely registered pair of images from which to extrapolate the stereo information; it works badly when the images contain several objects at very different distances from the viewer; polarised displays (which might solve the colour-washout problem) are only at the laboratory stage.[82]

Digital Video

Digital video can also take the viewer into an environment, and offers excellent quality which (another

---

[76] The live version is at http://vandyck.anu.edu.au/work/ multimedia.trials/popolo.tests/chigi/.

[77] As in this Vishnu in the National Gallery of Australia: http://rubens.anu.edu.au/new/presentations/sculpture/india/ national_gallery_of_australia/pala_vishnu/vishnumap/.

[78] Live hotspotted version at http://rubens.anu.edu.au/raid1/ cdroms/webready/england/oxford/cathedral/cathedral1/ wallpaper/wallpaper1.html.

[79] See Woods et al. (2000) for two decades of research.

[80] Warniers (2000); cf. Scharstein (1999), with details of 3D recognition and reconstruction techniques.

[81] That is, 3D laser imaging and processing: see http:// www.cee.hw.ac.uk/Vision/3D.html.

[82] Achieved by mounting two displays at right angles for the left and right eyes, one polarising vertically, the other horizontally, with a half-silvered mirror at 45°: 'This technique is capable of really exciting results' (Brice, 1997, 262).

Figure 5.6 Web imagemap: section of the Chigi Chapel, S. Maria del Popolo, Rome.

Figure 5.7 Web imagemap: sculpture of Vishnu in the National Gallery of Australia.

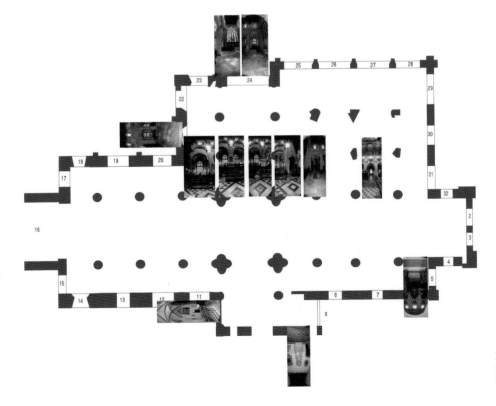

Figure 5.8 Web imagemap: 'wallpaper' nave section of Christ Church Cathedral, Oxford.

compromise) we must degrade in order to manage on today's computers.

Advantages: thanks to the spectacular zoom (10 to 20× without digital aid) of many digital video cameras, the medium can offer greater quality at less cost than any VRML model can currently achieve; the art historian can record the heights of church architecture or ceiling fresco much more easily than with any still camera, and hand-hold at very low light levels.

Disadvantages: digital video is jerky and bitty unless shot and edited to professional standards; it is of little use in cramped environments, because most cameras (like most digital still cameras) lack a sufficiently wide-angle lens; and one can trigger video from a hotspot, but cannot trigger computer actions of any kind from video. Digital video is unwieldy at any good resolution on any smallish computer, because of the huge quantities of data required.

All the above techniques have their value and drawbacks. Panoramas and stereo images are simple to construct but, except for the impressions of breadth in the former and depth in the latter, still keep the viewer behind the barrier of the photo plane (although zooming in and out offers some kind of autonomy). The concatenation of panoramas 'in depth' is more like a real-life experience, but still the user cannot manipulate what is seen — only move around the circular panoramas, in and out like some waltz figure. Getting lost is a real possibility, although location can be tracked on a plan.

Conclusion: Future Trends

Until there is a light, handheld device that can be used conveniently in an ill-lit church (no chariots with multiple video cameras; no turntables for laser recording; no heavy and specialised post-processing), then the following must form the modest but efficient panoply for the art historian wishing to recreate context:

1 if funds are available, VRML for a general (even exciting) overview — if funds are tight, maps or plans can equally well show the user the lie of the land; if hotspotted as imagemaps, the intricacies of the monument(s) under study can then be illustrated in an adjacent window;
2 ordinary panoramic images for a simply constructed wide view of the world;
3 vertical panoramas (which I call wallpaper, because they can cover whole walls) for a very simple way of laying out a monument, with associated imagemaps to offer ever-greater detail;
4 hotspotted or linked panoramas for more elaborate presentations;
5 stereo pairs, which can be of any size, and which can offer a remarkably effective impression of the third dimension.

A cool head is required when moving through the web-halls of Virtual Reality, because there is more rhetoric than in Byronic poetry, and more dead ends than in any episode of *Inspector Morse*: promise consistently outstrips delivery (remember *AI*? remember the Fifth Generation computer with which MITI and Japan would conquer the world?), and enthusiasms hooked to grandiose projects sometimes run into the dust as funding dries up. This may be a constant in some areas of computing, and it is dangerous for the timid art historian to be seduced by the magic box that is the computer (throw everything in, and the machine will deal with it), and by the exciting technologies available on the web, and as a result to mistake the appearance for the reality, and to jettison our simple scholarly requirements of accuracy and detail because of the compromises needed with what the software can do. Hence the software actually dictates what the project can achieve, whereas it should be the projected use which determines the necessary software. The initiative should always be with the user (I need software to do so-and-so) rather than with the software (this is a clever effect: what can we use it for?) if the modelling of the world on computers and across networks is to be a serious tool for learning and research.[83] This is not to condemn research and development in VR techniques, but simply to plead for a clear head in assessing their effectiveness. Hence, until convinced by flexible and cost-effective demonstrations on full-scale environments, we should concentrate on building applications which are simple, sustainable in their current state, and updatable in five years' time to take account of innovations. Web projects should, in this respect, try hard to imitate the longevity of the book.

If this discouraging picture of VR is accurate, what changes in the varieties of graphical exposition discussed above might we reasonably expect within five years? Should anyone today contemplating a sophisticated VRML project wait, or perhaps go ahead now with different ways of conjuring context? VRML may well remain in the realm of the generalised, so other avenues should be contemplated, especially if large-screen technologies develop apace (we show 35 mm slides on two large screens, so why not computer output?). The cathode ray tube monitor, unfortunately far from obsolete, is currently being challenged by thin-screen technology, which is widespread in the Japanese market. And if TFT technology will not in the immediate future be sufficiently cheap to offer wall-sized displays, we may nevertheless look forward to dual (front) video projection of computer images in the lecture theatre onto adjacent right-angled walls so that panoramas, imagemaps, and even stereo can take their place in the enhanced battery of graphical aids to scholarship as well as learning.

In summary, then, the would-be virtual reality neophyte might proceed according to four crushingly simple

---

83 See Hubbard (1995) for an early warning about who should be in control.

guidelines. These are frequently forgotten or evaded, so bear repeating here, notionally in flashing capital letters like a badly designed website. The first is that there is little point (other than convenience) in using the web simply to reproduce the two-screen darkened-room lecture set-up: there must be added value to make its use worthwhile, so that the levels of accuracy and detail offered by the examined technologies are not inferior to those attainable in ordinary photos or slides. The second is to know what one wishes to achieve, and then to examine the proposed software doing something very close to it, and to beware flashy demo programs. If no software does what you want, then ask very vigorously why not. The third is that VR technologies must be capable of implementation in a fashion that is cost- and time-efficient, sometimes by non-programmers. The fourth and final criterion is a reminder (somewhat akin to Mark 2:27) that digital/web is a medium, not a religion, and that, unlike Tertullian,[84] we should not believe simply because IT is impossible.

References

ADDISON, A. C. and GAIANI, M. (2000), 'Virtualized architectural heritage: new tools and techniques', *IEEE Multimedia* 7/2: 26–31.

ANTONIETTI, A. and CANTOIA, M. (2000), 'To see a painting versus to walk in a painting: an experiment on sense-making through virtual reality', *Computers & Education* 34/3–4: 213–23.

BARCELÓ, J. A., FORTE, M. and SANDERS, D. H., eds. (2000), *Virtual Reality in Archaeology*, BAR International Series 843 (Oxford).

BEISER, L. (1988), *Holographic Scanning* (New York).

BERNDT, E. and TEIXERA, J. C. (2000), 'Cultural heritage in the mature era of computer graphics', *IEEE Computer Graphics* 20/1: 36–7.

BRICE, R. (1997), *Multimedia and Virtual Reality Engineering* (Oxford).

BROOKS, F. P. (1999), 'What's real about virtual reality?', *IEEE Computer Graphics* 19/6: 16–27.

CAHEN, O. (1989), *L'image en relief, de la photographie stéréoscopique à la vidéo 3D* (Paris).

CHEN, C. (1999), *Information Visualization and Virtual Environments* (Berlin and London).

COMMENT, B. (2000), *The Painted Panorama* (New York).

CZERNUSZENKO, M., et al. (1999), 'Modelling 3D scenes from video', *The Visual Computer* 15: 341–8.

DAM, A. VAN, et al. (2000), 'Immersive VR for scientific visualization: a progress report', *IEEE Computer Graphics and Applications* 20/6: 26–52.

DEBEVEC, P. (2001), 'Pursuing reality with image-based modeling, rendering, and lighting', *3D Structure from Images: SMILE 2000: Second European Workshop on 3D Structure from Multiple Images of Large-Scale Environments, Dublin, Ireland, July 2000*, Lecture Notes in Computer Science 2018 (New York and London).

DELANEY, B. (2000), 'Visualization in urban planning: they didn't build LA in a day', *IEEE Computer Graphics* 20/3: 10–16.

DeLEON, V. J. (1999), 'VRND: Notre-Dame Cathedral: a globally accessible multi-user real-time virtual reconstruction', *Proceedings of the International Conference on Virtual Systems and Multimedia VSMM '99* (Dundee).

DONG, W. and GIBON, K. (1998), *Computer Visualisation: an Integrated Approach for Interior Design and Architecture* (New York).

DURLACH, N. I. and MAVOR, A. S., eds. (1995), *Virtual Reality: Scientific and Technological Challenges*, Committee on Virtual Reality Research and Development (Washington, DC).

FRISCHER, B., et al. (2000), 'Virtual reality and ancient Rome: the UCLA Cultural VR Lab's Santa Maria Maggiore Project', in J. A. Barceló et al. (eds.), *Virtual Reality in Archaeology*, BAR International Series 843 (Oxford): 155–62.

FUNKHAUSER, T. and LI, K. (2000), 'Large format displays', *IEEE Computer Graphics and Applications*, July–August: 20–1.

FURHT, B., ed. (1999), *Handbook of Multimedia Computing* (Boca Raton).

GREENHALGH, C. (1999), *Large-Scale Collaborative Virtual Environments* (London).

HALL, W., WHITE, S. and WOOLF, B. P. (1999), 'Interactive systems for teaching and learning', in B. Furht (ed.), *Handbook of Internet and Multimedia Systems and Applications* (Boca Raton): 375–97.

HAVAL, N. (2000), 'Three-dimensional documentation of complex heritage structures', *IEEE Multimedia* 7/2: 52–5.

HEIM, M. (1998), *Virtual Realism* (New York and Oxford).

HUBBARD, D. (1995), 'Virtuality and rumors of reality: the humanist in an interactive age', *College Language Association Journal* 39: 1–17.

IEEE COMPUTER GRAPHICS (1997), 17/2 [dedicated to 3D and multimedia on the information superhighway].

IEEE COMPUTER GRAPHICS (1999a), 19/2 [dedicated to VRML].

IEEE COMPUTER GRAPHICS (1999b), 19/6 [dedicated to virtual reality].

IEEE COMPUTER GRAPHICS (2000), 20/2 [dedicated to perceiving 3D shape].

JIN, L. and WEN, Z. (2001), 'Adorning VRML worlds with environmental aspects', *IEEE Computer Graphics and Applications* 21/1: 6–9.

KENDERDINE, S. (1998), 'Sailing on the silicon sea: the design of a virtual maritime museum', *Archives & Museum Informatics* 12/1: 17–38.

KRANSTEUBER, A. S. (2000), 'Real time holographic display and image processing: a dissertation', PhD typescript (University of Alabama at Huntsville).

KRISZAT, M. (1999), 'Hypermedia in archaeological exhibitions: different kinds of interactivity and visualization', *Archives & Museum Informatics* 13/2: 139–67.

[84] Tertullian, *De Carne Christi* v.

KURBEL, K. and TWARDOCH, A. (2000), 'State-of-the-art multimedia technologies for website construction', *Wirtschaftsinformatik* 42/3: 253–64.

LEVOY, M. et al. (2000), 'The digital Michelangelo project: 3D scanning of large statues', *Siggraph 2000: Computer Graphics Proceedings*, July: 131–44.

LUDLOW, J. B. and PLATIN, E. (2000), 'A comparison of web page and slide/tape for instruction in peri-apical and panoramic radiographic anatomy', *Journal of Dental Education* 64/4: 269–75.

MANGAN, K. A. (2000), 'Teaching surgery without a patient', *Chronicle of Higher Education* 46/25: A49–50.

MARTY, P. F. (2000), 'On-line exhibit design: the socio-technological impact of building a museum over the World Wide Web', *Journal of the American Society for Information Science* 51/1: 24–32.

MEI, Q. and WING, R. (1999), 'Robotic 360 degrees photography for virtual site visits', in E. Banissi et al. (eds.), *1999 IEEE International Conference on Information Visualization* (Los Alamitos, CA): 214–19.

MOHLER, J. L. (2000), 'Creating a web-based virtual tour at Purdue University', *Journal of Computing in Higher Education* 11/2: 50–62.

MÜLLER, W. and QUIEN, N. (1999), *Spätgotik Virtuell: für und wider der Simulation historischer Architektur* (Weimar).

NAKAMAE, E. et al. (1999), 'Computer generated still images composited with panned/zoomed landscape video sequences', *The Visual Computer* 15: 429–42.

NOVITSKI, B. J. (1998), 'Reconstructing lost architecture', *Computer Graphics World* 21/12: 24–30.

NOVITSKI, B. J. (1999), *Rendering Real and Imagined Buildings: The Art of Computer Modeling from the Palace of Kublai Khan to Le Corbusier's Villas* (Gloucester, MA).

OETTERMANN, S. (1997), *The Panorama: History of a Mass Movement* (New York).

PAOLINI, P., BARBIERI, T., LOIUDICE, P., ALONZO, F., ZANTI, M. and GAIA, G. (2000), 'Visiting a museum together: how to share a visit to a virtual world', *Journal of the American Society for Information Science* 51/1: 33–8.

PAPAIOANNOU, G., KARABASSI, E.-A. and THEOHARIS, T. (2001), 'Virtual archaeologist: assembling the past', *IEEE Computer Graphics and Applications*, March–April: 53–9.

PELEG, S. and BEN-EZRA, M. (1999), 'Stereo panorama with a single camera', *1999 IEEE Computer Society Conference on Computer Vision and Pattern Recognition*, vol. 1: 395–401.

PELEG, S., BEN-EZRA, M. and PRITCH, Y. (2001), 'Omnistereo: panoramic stereo imaging', *IEEE Transactions on Pattern Analysis & Machine Intelligence* 23/3: 279–90.

PERETZ, L. (1989), *L'image en 3 dimensions* (Paris).

PLANTE, A., TANAKA, S. and IWADATE, Y. (1999), 'Virtual Shinto shrine', *Proceedings of the IEEE International Conference on Multimedia Computing and Systems*, vol. 2: 1024–5.

PORTER, T. and NEALE, J. (2000), *Architectural Supermodels: Physical Design Simulation* (Oxford).

RICE, S. (1993), 'Boundless horizons: the panoramic image. Exhibition, Bonn: Sehsucht; das Panorama als Massenunterhaltung des 19. Jahrhunderts', *Art in America* 81, Dec.: 68–71.

SCHARSTEIN, D. (1999), 'Re-evaluating stereo', *View Synthesis Using Stereo Vision*, Lecture Notes in Computer Science 1583 (Berlin and London): 63–74.

SCHUBERT, R. (2000), 'Using a flatbed scanner as a stereoscopic near-field camera', *IEEE Computer Graphics* 20/2: 38–45.

SHULMAN, S. (1998), 'Digital antiquities', *Computer Graphics World* 21/11: 34–8.

SHUM, H. Y. and SZELISKI, R. (2000), 'Construction of panoramic image mosaics with global and local alignment', *International Journal of Computer Vision* 36/2: 101–30.

SINGH, G., et al., eds. (1994), *Virtual Reality Software and Technology: Proceedings of the VRST 94 Conference* (Singapore).

VVECC (2000), *Augmenting the Real World: Augmented Reality and Wearable Computing* [Visualization and Virtual Environments Community Club] http://www.avrrc.lboro.ac.uk/vveccar.html.

WARNIERS, R. (1998a), 'Dirty pictures', *Computer Graphics World* 21/6: 50–60.

WARNIERS, R. (1998b), 'Every picture tells a story', *Computer Graphics World* 21/10: 25–32.

WARNIERS, R. (2000), 'Picture-perfect modelling', *Computer Graphics World* 23/6: 54–58.

WEI, S. K., HUANG, Y. F. and KLETTE, R. (1999), 'Three-dimensional scene navigation through anaglyphic stereo', *Computer Analysis of Images and Patterns: 8th International Conference, CAIP '99 Proceedings*, Lecture Notes in Computer Science 1689: 542–9.

WOODS, A. J., et al., eds. (2000), *Selected SPIE papers on CD-ROM: Stereoscopic Displays and Applications: Proceedings of the Stereoscopic Displays and Applications Conference* (Washington) [2 CD-ROMs].

XIAO, D. Y. (2000), 'Experiencing the library in a panorama virtual reality environment', *Library Hi Tech*, 18/2: 177–84.

YAMADA, K., ICHIKAWA, T., NAEMURA, T., AIZAWA, K. and SAITO, T. (2000), 'High-quality stereo panorama generation using a three-camera system', *Proceedings of SPIE, the International Society for Optical Engineering* 4067/1–3: 419–28.

YOUNG, J. R. (2000), 'Virtual reality on a desktop hailed as new tool in distance education', *Chronicle of Higher Education* 47/6: A43–4.

CHAPTER 6

Automatic Creation of Virtual Artefacts from Video Sequences

Andrew W. Fitzgibbon, Geoff Cross & Andrew Zisserman

I. Introduction

In order to measure, analyse, or simply admire artefacts which are located far from the user, it is useful to have a digital representation of the artefact. For the purpose of recording, storing, or transmitting such information, digital photographs suffice. However, for examination of the artefact, a 3D presentation is invaluable, allowing the object viewpoint to be modified freely, and 3D measurements to be taken on object features. In this paper we describe a system which can acquire such 3D models from photographs, without the need for calibration of system geometry such as the camera focal length, relative positions of the camera and object, or relative motion of the camera and object. The system instead computes a representation of all possible object and camera configurations which are consistent with the given images.

The work is presented in three parts. The first describes how tracking of points observed in two-dimensional images permits the computation of the relative camera and object geometry. Second, we describe the construction of a triangulated 3D model from the object silhouettes. Finally, the refinement of the model based on surface texture is discussed, and examples of the system's performance are provided.

II. Recovery of Camera Motion and Scene Geometry

The first concern in acquiring a 3D model is the determination of the relative geometry of camera and scene.

By tracking 2D features through a sequence, a general structure and motion recovery algorithm can compute 3D camera motion, 3D co-ordinates of the points, and the camera's internal calibration parameters. This work treats a common special case of the 3D reconstruction problem, where the camera motion is restricted to rotation about a single world axis. The standard example of this configuration is an object on a rotating turntable, observed by a single static camera.[1] It also occurs when a boom-mounted camera is rotated,[2] and when a camera

is on a tripod and the optical centre is not coincident with the rotation axis. Several example sequences are shown in Figure 6.1.

Several benefits accrue from considering this arrangement: a direct (non-iterative) n-view solution can be found, which has not yet been possible for the general problem; accuracy is increased because fewer parameters need to be estimated; and robustness is improved by the simultaneous batch analysis of all the data in the sequence. The experiments reported in the paper confirm these expectations, with reconstruction accuracies which are an order of magnitude better than the best general motion results.

We begin by reviewing previous approaches to single-axis reconstruction and related problems. Then the geometry of the uncalibrated case is developed. The reconstruction ambiguity in the uncalibrated case is shown to correspond to a two-parameter projective ambiguity along the axis of rotation. The rotation angles are unambiguously computed when more than two views are available. Supplying a modicum of geometric information, or camera calibration, results in a unique solution. The geometric analysis is completed by deriving the form of the fundamental matrix. Section III then describes the estimation of the fundamental matrix from point tracks, and hence the recovery of the entire system geometry. This estimate is refined by nonlinearly fitting all tracks to the single-axis model using bundle adjustment.

(a) Background: Simplifications of the General Structure and Motion Problem

The recovery of structure and motion from point tracks has been widely studied in recent years.[3] Of particular interest in this paper is the work that has been performed on special cases of the general motion recovery problem. Most of these endeavours have been characterised by the simplifications which they confer upon the problem, thereby rendering difficult cases tractable.

The first class of simplifications arise from the use of approximations to the 'true' pinhole camera model. Early work on structure and motion recovery using the orthographic camera[4] allowed closed-form solutions to

---

[1] Sawhney et al. (1990); Szeliski (1991); Szeliski and Kang (1993); Niem and Buschmann (1995); Kang (1997).
[2] Shum and He (1999).

[3] Hartley and Zisserman (2000).
[4] Harris (1990); Tomasi and Kanade (1992).

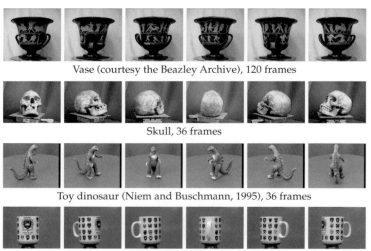

Vase (courtesy the Beazley Archive), 120 frames

Skull, 36 frames

Toy dinosaur (Niem and Buschmann, 1995), 36 frames

Cup, 36 frames

Figure 6.1 Some example sequences. In each case, camera calibration is unknown, and turntable rotation angles are unknown.

be found. Similarly, the extension to the affine camera[5] allows the closed-form estimation of multiple cameras and 3D point positions. However, under the imaging conditions which typically apply in close-range model acquisition, this approximation can be quite poor, even though the pinhole camera model is itself quite accurate (i.e. non-projective lens distortion is small).

In the second class of simplifications the pinhole projection model is not approximated; however, the type of motion the camera can undergo is restricted. Specialising to pure translation[6] or planar motion[7] simplifies autocalibration and increases robustness when the modelled motion is correct. Importantly, such constraints are more readily satisfied in typical applications than is the placement of cameras at a great distance from the scene.

The case studied here is in fact itself a special case of planar motion that is particularly easy to arrange — namely, objects which are viewed on a rotating turntable. This scenario has been studied in the past[8] but the special motion has not previously been exploited to simplify the camera recovery.

This paper advances the field by considering the uncalibrated case of single-axis motion. We characterise all solutions to reconstruction from point tracks, and show that a two-parameter family of reconstructions exists, all of which predict the 2D point tracks equally well. We show that when further information is available, for example regarding the camera, the two-parameter ambiguity can be eliminated. The material presented here has previously appeared in abbreviated form,[9] and has been developed further elsewhere.[10]

(b) The Geometry of Single-Axis Rotation

It will be helpful in the following to consider that the object is fixed and that the camera rotates about it. In addition, we may sometimes think of the situation as being several static cameras viewing a static scene. The camera's internal parameters are fixed. To aid visualisation, we assume that the rotation axis is vertical, so that the camera rotates in a horizontal plane.

This analysis deals with the geometry of 3D point sets when viewed as 2D points under perspective projection.[11] It is assumed that 2D tracks are provided — the image co-ordinates of a 3D point in two or more views. Thus the scene is a collection of n 3D points $\{\mathbf{X}_j\}_{j=1}^{n}$. What is observed is the set of 2D images x_{ij} of each point $\mathbf{X}_j$ in some subset of the m views. The camera position (calibration, orientation, and translation) with respect to the 3D points is represented by a 3×4 projection matrix $\mathtt{P}_i$; 3D points $\mathbf{X}$ in the scene are represented as homogeneous 4-vectors $[x, y, z, 1]^{\top}$, while their 2D projections x are represented as homogeneous 3-vectors $[x, y, 1]^{\top}$. The action of each camera is represented by

$$x_{ij} = \mathtt{P}_i \mathbf{X}_j$$

The m cameras are indicated by $\mathtt{P}_i$ while the n 3D points are $\mathbf{X}_j$.

In the case where m different cameras view a scene, there is no relationship between the $\mathtt{P}_i$. Therefore $11m$ parameters are required to specify all the cameras. When the cameras have identical internal parameters, such as when a camera is moved through a static scene without any change in focus or zoom, the internal parameters are constant over the sequence, so that only rotation and translation (6 parameters) need be parametrised in each view. This reduces the number of parameters required to specify the cameras from $11m$ to $6m + 5$. In the single-axis case, we shall see that the number of parameters is reduced to $m + 8$.

[5] Koenderink and Van Doorn (1991); Demey et al. (1992); Mundy and Zisserman (1992); Quan and Mohr (1992); Poelman and Kanade (1994); Shapiro et al. (1995).

[6] Moons et al. (1994).

[7] Beardsley and Zisserman (1995); Viéville and Lingrand (1995); Armstrong et al. (1996).

[8] Sawhney et al. (1990); Szeliski (1991); Szeliski and Kang (1993); Kang (1997).

[9] Fitzgibbon et al. (1998); Cross and Zisserman (2000).

[10] Jiang et al. (2001); Mendonça et al. (2001).

[11] Hartley and Zisserman (2000).

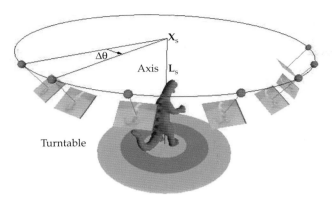

Figure 6.2 3D geometry. The cameras are indicated by their centres (spheres) and image planes. The point X_s is the intersection of the plane π_h containing the camera centres with the rotation axis L_s.

The geometry of single-axis rotation is illustrated in Figure 6.2. Because we have the freedom to choose co-ordinates, we may define the rotation axis to be the world z axis. The camera centre describes a circle in a plane perpendicular to this axis, which we choose to coincide with the xy plane.

In order to derive the special form of the camera projection matrices P for single-axis rotation, we consider a camera in an arbitrary position

$$P = K[R| -Rt]$$

where R is a 3×3 rotation matrix representing the camera orientation, t is the position of the optical centre, and K is the upper-triangular calibration matrix:

$$K = \begin{bmatrix} f & s & u_0 \\ 0 & af & v_0 \\ 0 & 0 & 1 \end{bmatrix} \begin{cases} f \text{ is focal length} \\ (u_0, v_0) \text{ is the principal point} \\ a \text{ is aspect ratio; } s \text{ is skew} \end{cases}$$

Thus the first camera may be written

$$P_0 = H[I | -t]$$

where $H = KR$ is a homography representing the camera's internal parameters and rotation about the camera centre. We have the further freedom in the world co-ordinate system to choose the first camera centre to be the x axis and to choose the overall scale, so we set $t = (-1, 0, 0)^\top$.

A rotation of the camera by θ about the z axis is achieved by post-multiplying P_0 by a 4×4 homogeneous transformation matrix corresponding to rotation about z

$$\begin{bmatrix} R_Z(\theta) & \mathbf{0} \\ \mathbf{0}^\top & 1 \end{bmatrix}$$

yielding the camera $P_\theta = H[R_Z(\theta) | -t]$. In detail, with h_i the columns of H:

$$P_\theta = \begin{bmatrix} h_1 & h_2 & h_3 \end{bmatrix} \begin{bmatrix} \cos\theta & \sin\theta & 0 & 1 \\ -\sin\theta & \cos\theta & 0 & 0 \\ 0 & 0 & 1 & 0 \end{bmatrix} \quad (6.1)$$

This division of the internal and external parameters means that H and t are fixed over the sequence; only the angle of rotation about the z axis, θ_i, varies for each

camera P_i. Given this parametrisation, the estimation problem can now be precisely stated: we seek the common matrix H and the angles θ_i in order to estimate the set of cameras P_i for the sequence. Thus a total of $8 + m$ parameters must be estimated for m views, where 8 is the number of degrees of freedom of the homography H. The relative angle between views i and $i + 1$ is denoted $\Delta\theta_i$.

(c) The Fundamental Matrix in Terms of H and θ

The previous discussion has given the form of the camera projection matrices. In the recovery of general structure and motion,[12] an effective strategy is to compute the fundamental matrix for each pair of views, and then extract the parameters of the 3D geometry from the fundamental matrix.

We would like to know the form of the fundamental matrix in terms of the unknowns of the system geometry, specifically the columns of the homography $h_{1\ldots3}$, and the rotation angles θ_i. Without loss of generality, we may assume the two views are P_0 and P_θ. Let us temporarily transform image co-ordinates by the (as yet unknown) homography H, writing $\hat{x} = H^{-1}x$. Calling the fundamental matrix in this co-ordinate system $\hat{F}$, we have $F = (H^{-\top}\hat{F}H^{-1})$ and

$$\hat{P}_0 = [I | -t]$$
$$\hat{P}_\theta = [R_Z(\theta) | -t]$$

Because we are using these matrices to compute $\hat{F}$, which is invariant to transformations of world co-ordinates, we may post-multiply both camera matrices by the 3D transformation

$$\begin{bmatrix} I & t \\ \mathbf{0}^\top & 1 \end{bmatrix}$$

to give the canonical forms

$$\hat{P}_0 = [I | \mathbf{0}]$$
$$\hat{P}_\theta = [R_Z(\theta) | -t + R_Z(\theta)t]$$

from which the fundamental matrix $\hat{F}$ may be written

$$\hat{F} = [-t + R_Z(\theta)t]_\times R_Z(\theta) = \begin{bmatrix} 1-c \\ s \\ 0 \end{bmatrix}_\times \begin{bmatrix} c & s & 0 \\ -s & c & 0 \\ 0 & 0 & 1 \end{bmatrix}$$

$$= \begin{bmatrix} 0 & 0 & s \\ 0 & 0 & c-1 \\ -s & c-1 & 0 \end{bmatrix}$$

where $c = \cos\theta$ and $s = \sin\theta$, and the 3×3 antisymmetric matrix $[v]_\times$ is such that $[v]_\times x = v \times x \ \forall x$.

Decomposing into symmetric and antisymmetric parts, we obtain

$$\hat{F} = s \begin{bmatrix} 0 & 0 & 1 \\ 0 & 0 & 0 \\ -1 & 0 & 0 \end{bmatrix} + (c-1) \begin{bmatrix} 0 & 0 & 0 \\ 0 & 0 & 1 \\ 0 & 1 & 0 \end{bmatrix}$$

[12] Faugeras (1992); Hartley et al. (1992); Beardsley et al. (1996); Fitzgibbon and Zisserman (1998).

and

$$F = H^{-\top}\hat{F}H^{-1} = sH^{-\top}\begin{bmatrix} 0 & 0 & 1 \\ 0 & 0 & 0 \\ -1 & 0 & 0 \end{bmatrix}H^{-1}$$

$$+ (c-1)H^{-\top}\begin{bmatrix} 0 & 0 & 0 \\ 0 & 0 & 1 \\ 0 & 1 & 0 \end{bmatrix}H^{-1} \quad (6.2)$$

Some fairly sprightly algebraic manipulation allows us to express this in terms of the columns of H, giving

$$F = [h_2]_\times - \frac{1}{(\det H)} \tan\frac{\Delta\theta}{2}\left((h_1 \times h_3)(h_1 \times h_2)^\top\right.$$

$$\left. +(h_1 \times h_2)(h_1 \times h_3)^\top\right) \quad (6.3)$$

It is this form of the fundamental matrix that we shall fit to the point tracks, and thence compute the set of consistent homographies H. The matrix has rank 2, in common with the general motion fundamental matrix, but satisfies the additional constraint that its symmetric part $F + F^\top$ also has rank 2.

As noted above, reconstruction from uncalibrated cameras will in general yield a family of models all of which generate the same image point tracks. Figure 6.3 illustrates the range of models which can generate the images in the 'cup' sequence. In the absence of further information, we cannot choose between these solutions. On the other hand, if some camera information is available, Section III(d) shows how the metric structure can be recovered *post hoc*.

The reconstruction ambiguity is the following:[13] metric structure is recovered in planes perpendicular to the axis of rotation; there is an unknown 1D projective transformation along the axis. The ambiguity may be written:

$$\begin{pmatrix} x \\ y \\ z \end{pmatrix} \rightarrow \begin{pmatrix} x/(\beta z + 1) \\ y/(\beta z + 1) \\ \alpha z/(\beta z + 1) \end{pmatrix}$$

Note that since metric structure is determined in planes perpendicular to the axis, the angle of rotation between views is known.

Figure 6.3 Projective ambiguity. With no information about the camera or scene, there is a 1D projective ambiguity in the z direction. Any of the five models of the cup with different choices for H admits a corresponding camera geometry which can explain the images in Figure 6.1(d). The ambiguity is removed by e.g. specifying the aspect ratio and identifying a pair of parallel lines in the scene, or taking calliper measurements of a pair of object points which are identifiable in the images.

[13] Zisserman et al. (1995).

III. Estimation of Geometry

In this section we assume that we are given 2D point tracks, and show how to extract the system geometry from these data. The extraction is not sensitive to missing data, and does not require that any point be visible in more than two views.

(a) Computing the Fundamental Matrix

Section II(b) shows that the fundamental matrix for single-axis rotation has 6 degrees of freedom. As the matrix has rank 2, estimation of the single-axis rotation F is via a modification of the standard 7-point algorithm.[14] The general motion fundamental matrix is computed, and decomposed into symmetric and antisymmetric parts. The eigensystem of the symmetric part is computed and the eigenvalue closest to zero is set to zero. Recomposing the symmetric part and adding the antisymmetric part returns a matrix of the appropriate form. Pseudocode is shown in Figure 6.4.

(b) Decompositon of the Fundamental Matrix

The computational strategy will be to compute the fundamental matrix from point correspondences, and then to extract H from F. This section derives an algorithm for this decomposition. Extraction of h_2 is straightforward: compute the antisymmetric part, $\frac{1}{2}(F - F^\top)$, and extract the three components.

Extraction of the remaining columns of H follows from Equation (6.3). The first column h_1 is obtained as the null space of $F + F^\top$. Column h_3 is determined as follows:

1. Set $l_h = h_1 \times h_2$
2. Set $l_s = (2l_h^\top l_h I - l_h l_h^\top)(F + F^\top)l_h$
3. Set $h_3 = l_s \times (0, 0, 1)^\top$

This selects one member from the family of homographies consistent with the given fundamental matrix. Calling this matrix H_0, the remaining members of the family are given by

$$H = H_0 \begin{pmatrix} 1 & 0 & \nu \\ 0 & \mu & 0 \\ 0 & 0 & \omega \end{pmatrix}$$

where ν and ω parametrise the two-parameter family of solutions and μ cannot be determined until $\Delta\theta$ is known.

1. Compute symmetric part $S = \frac{1}{2}(F+F^\top)$
2. Compute eigensystem $S = VDV^\top$
3. Set $D_{ii} = 0$ where $i = \text{argmin}_i \|D_{ii}\|$
4. Recompute $F_{sar} = \frac{1}{2}(F-F^\top) + VDV^\top$

Figure 6.4 Pseudocode for truncation of general motion fundamental matrix to single-axis form.

[14] Torr and Murray (1997).

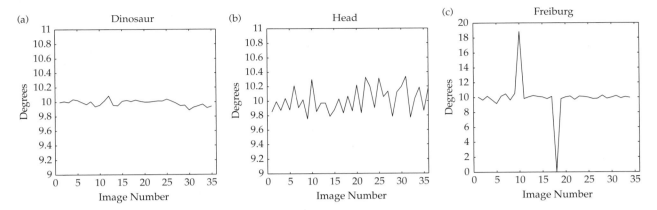

Figure 6.5 Geometry estimation. The graphs show the recovered angles between successive views for each of three sequences. (a) Object rotated by a mechanical turntable with a resolution of 1 millidegree. The RMS difference between the angle recovered by our algorithm and the nominal value is 40 millidegrees. This demonstrates the accuracy of the angle recovery. (b), (c) Turntable rotated by hand. The angle increment is irregular and unknown a priori. Variation is up to 20° because of missing and repeated views.

(c) Nonlinear Refinement

The two-view geometry provides an initial estimate for the camera matrices. Estimates for the turntable angles and the μ_i are obtained by averaging single-track estimates over three views.[15] In order to determine the maximum likelihood estimate, we assume that errors in the positions of the 2D points are normally distributed. An optimal estimate is then obtained by nonlinear minimisation of the distances between the reprojected 3D points and the 2D corners.[16] Typical results for geometry estimation are shown in Figure 6.5. These results are of comparable quality with those of Szeliski and Kang (1993) where the camera matrices were determined using a calibration pattern. Convergence is generally achieved in 8 iterations, reducing the RMS reprojection error from 0.3 to 0.1 pixels. For 2000 points, compute time per iteration is of the order of 10 seconds on a 300 MHz UltraSparc. The radius of convergence is large, the correct minimum being achieved from initial estimates where the θ_i are in error by up to a factor of 2, although of course many more iterations (about 100) are required.

(d) Removing the Reconstruction Ambiguity

To this point no information on the internal calibration of the camera, or on the 3D shape of the object, has been used. Internal constraints are provided, for example, by the fact that the image pixels have zero skew, and known aspect ratio. Often the zero skew constraint is not useful in practice because it does not resolve the ambiguity.[17] For example, if the image plane is parallel to the rotation axis then all members in the family of solutions for the calibration matrix will already satisfy the zero-skew constraint.

It can be shown that specifying the aspect ratio places a quartic constraint on the parameters α, β.

The easiest method of resolution is to use a pair of parallel verticals in the scene to identify the vanishing point at infinity in the z direction. This determines h_3 up to scale (i.e. the ratio $\alpha : \beta$), and the only remaining ambiguity is then a relative scaling of the z and plane directions:

$$\begin{pmatrix} x \\ y \\ z \end{pmatrix} \rightarrow \begin{pmatrix} x \\ y \\ \alpha' z \end{pmatrix}$$

Given h_3 up to scale, the internal aspect ratio then determines α and β uniquely, up to an arbitrary choice of sign. Alternatively, the aspect ratio of the object can be used to resolve the ambiguity.

IV. Results: Estimation

Evaluation of the geometry estimation was based on 2D point tracks automatically obtained by interest-point detection[18] followed by two-view matching and multiple-hypothesis tracking.[19] On a typical sequence (the dinosaur sequence in Figure 6.1), 500 points are detected per image. Tracking then yielded an average of 340 tracks between each pair of successive frames, for a total of about 5000 tracks. The mean track length was 3.3 frames, with 37 tracks longer than 10 frames. No track lived for longer than 20 frames. The track lifetimes are illustrated in Figure 6.6, where it can be seen that, although numerous, the tracks are not densely sampled through the sequence.

Camera geometry was computed using the algorithm described above, and the results are illustrated in Figure 6.7. Quantitative results are shown in Figure 6.5. On the sequence for which ground truth is available, the

[15] Fitzgibbon et al. (1998).
[16] Hartley (1994).
[17] Zisserman et al. (1998).

[18] Harris (1990).
[19] Torr et al. (1999).

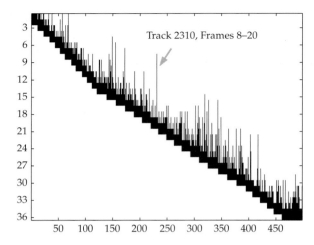

Figure 6.6 Track lifetimes. The y-axis corresponds to frame numbers on the dinosaur sequence. Each column represents a single 2D track, and indicates the frames of the sequence in which that track was seen.

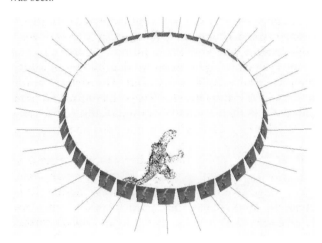

Figure 6.7 Dinosaur: points and cameras reconstructed from 2D point tracks.

(a)   (b)

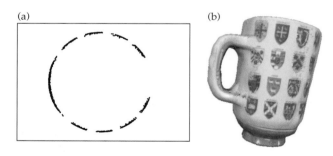

Figure 6.8 (a) Top view of reconstructed cup points (no points were detected on the handle). RMS difference from a fitted cylinder is 0.004 of the diameter. (b) Texture-mapped 3D reconstruction.

inter-view angles are estimated to an RMS accuracy of 0.04°. Figure 6.8 shows results of the 3D geometry estimation on an object with circular cross-section. In this case, an estimate of the accuracy of the system comes from fitting a cylinder to the recovered 3D points and measuring the deviation of the points from the surface. In this case, the RMS difference is 0.004 of the cylinder diameter, corresponding to about 0.4 mm for a 10 cm object size.

Figure 6.9 The volumetric constraint offered by a single silhouette. If the camera centre and image plane geometry is known, the 3D object must lie within the polyhedral cone containing the optical centre and the (polygon defining the) silhouette. Here the cone has been truncated at near and far planes outside which the object is known not to lie.

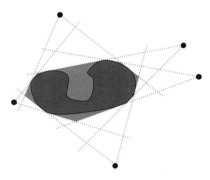

Figure 6.10 A 2D surface (dark shading) is viewed by five 1D cameras (•), and the visual hull is represented by light shading. Note that the concavity is never represented on any of the silhouettes and hence cannot be reconstructed from silhouette information alone: the visual hull cannot include this region.

Figure 6.11 The visual hull from the dinosaur sequence in Figure 6.1, reconstructed from (left to right) 2, 4, and all 36 silhouettes. Note the natural increase in accuracy as the number of silhouettes increases.

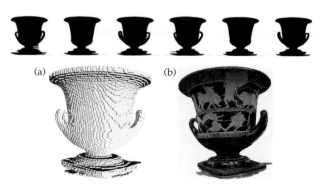

(a)   (b)

Figure 6.12 Six silhouette images from a sequence of 120 images of a vase. The silhouette is badly segmented (note around the base in the first and fourth image, for example), but the visual hull (a) is accurately represented by the voxmap. (b) The vase reconstructed and texture-mapped. All 120 images are used in the reconstruction, and the voxmap is generated at a resolution of 512^3.

V. Reconstruction from Silhouettes, Known Camera Motion

We now consider the construction of 3D solid models by volume intersection from multiple views. As pointed out by Sullivan and Ponce (1998), the idea dates back to Baumgart (1974). Well-engineered systems built on this idea have yielded 3D texture-mapped graphical models of impressive quality.[20] A good example is the system of Niem and Buschmann (1995) where, as is usual for such systems, the object is rotated on a turntable against a background which can easily be removed by image segmentation. Volume intersection is an exacting test of the accuracy to which system geometry is known, and such systems have until now been completely pre-calibrated. In this section, the automatically computed geometry described above is used to create the models.

The basic principle of modelling from silhouettes is illustrated in Figure 6.9. A silhouette seen in a single view defines a generalised cone with apex at the camera centre, and whose surface is exactly the set of lines which pass through the camera centre and the points of the silhouette on the (3D) image plane. The object volume is clearly a subset of the interior of the cone. Given many such cones, areas of 3D space are virtually carved away, gaining a closer approximation to the object volume with each new cone. It is immediately evident, however, that with a finite number of views, the region so obtained will always be a superset of the true object volume. Even with an infinite number of views, concavities in the object cannot be seen, and will therefore not be modelled. In this case, the envelope of all possible cones is called the *visual hull*.[21] Figure 6.10 illustrates these concepts for a two-dimensional world and one-dimensional cameras. In 2D it can be shown that an object's visual hull and convex hull are coincident, but in 3D they are quite distinct, as the constraint obtained from even one view need not be convex.

In the next section we shall see how these difficulties are overcome. However, even a small number of silhouettes do provide a surprisingly accurate model of the object, as can be seen in Figures 6.11 and 6.12. This will prove a valuable starting point for the refinement strategy.

(a) Combined Texture and Silhouette Refinement

It has been shown that the apparent contour does not provide any information about concavities of the viewed surface. In order to accurately reconstruct such surface regions, it is necessary to make use of other information such as texture. Dense stereo algorithms make good use of texture information. The work of Pollefeys et al. (1998) represents one example of the state of the art. More recent efforts have followed the 'Space Carving' paradigm of Seitz and Dyer (1997) and Kutulakos and Seitz (1998), where a Monge patch or polygon-based

model is replaced by a volumetric occupancy grid, and careful management of the order in which voxels are processed means that hidden-surface computations are efficiently performed. For refinement of these models, efficient implementations are exemplified by the level set approaches of Faugeras and Keriven (1998). However, these texture-based strategies are prone to two significant failings:

1. Deviations from the matte surface assumptions which are implicit in most such implementations can cause 3D artefacts to appear on the surfaces, which do not occur in the extraction of 3D from silhouettes.
2. For stability, it is necessary to apply the consistency tests over relatively large image regions, meaning that fine details are lost. This is exacerbated if the matte reflectance assumptions are relaxed.

In this paper we combine the silhouette and texture-based constraints by eroding inconsistent surface voxels, to obtain accurate reconstructions.

The space carving technique uses a discrete volumetric representation of 3D space, and tests each voxel for photometric consistency when projected into the original images. Suppose that a certain voxel's centre is given by the 3D point $\mathbf{X}$, and its projection into two images is given by $\mathbf{x}_1 = P_1\mathbf{X}$ and $\mathbf{x}_2 = P_2\mathbf{X}$.

If the surface obeys the Lambertian reflectance law, its intensity will be the same from all viewing directions, so the image intensities will be the same modulo imaging noise. Thus, one matching criterion simply matches the image intensities between the hypothesised correspondences. That is, given an image point in each of two images, $\mathbf{x}_1$ and $\mathbf{x}_2$, with corresponding image intensity values given by $I_1(\mathbf{x}_1)$ and $I_2(\mathbf{x}_2)$, the matching score is typically defined as

$$\epsilon = (I_1(\mathbf{x}_1) - I_2(\mathbf{x}_2))^2 \qquad (6.4)$$

A low score indicates a good match. In most practical cases, this matching score is insufficient. Image noise — on $I_1(\mathbf{x_1})$ and $I_2(\mathbf{x_2})$ — affects ϵ, as does localisation error on $\mathbf{x}_1$ and $\mathbf{x}_2$. Further, ϵ is not robust to lighting changes between images, which generally result in an unknown transformation in intensities over the entire image. Using a normalised cross-correlation improves over that given by Equation (6.4). It is robust to small errors in $\mathbf{x}_1$ and $\mathbf{x}_2$ as the signal is locally correlated, and image noise is averaged over the texture patch. Further, a normalised correlation is invariant to lighting changes which induce a locally affine transformation on the image intensities. Figure 6.13(b) illustrates the computation of this metric on each voxel on the visual hull extracted from the skull silhouettes. Removing voxels with low correlation scores allows concavities on the object to be modelled. Repeating the computation until all surface voxels are texture-consistent yields the model shown in Figure 6.13(d).

---

[20] Boyer (1996); Sullivan and Ponce (1998).
[21] Laurentini (1994).

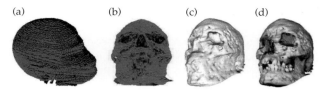

(a)   (b)   (c)   (d)

Figure 6.13 (a) A voxellated model of the visual hull of a skull. It can be seen that the visual hull does not include concavities such as the nose and eye sockets. (b) A photo-consistency function is applied to the surface of the visual hull, and regions with a high score are shown in red and those with a low score are shown in black. A high score indicates that the reprojection of the surface is consistent with the input images. It can be seen that the eyes and nose regions have been marked as 'incorrect'. (c) The final model after eroding inconsistent voxels; areas such as the eye socket and nose region are correctly reconstructed. A simple mesh-smoothing algorithm has been applied to the surface to highlight these areas. (d) A textured-mapped model using intensities from the original images.

VI. Discussion

A system which allows the construction of 3D models of artefacts rotated on a turntable and viewed by an a priori uncalibrated camera has been presented. The system uses the 2D tracks of textured points on the object as the turntable rotates to calibrate the system, determining the camera geometry and turntable angles up to a two-parameter ambiguity. The accuracy of the reconstruction is enhanced by using more points on the object, and is comparable to traditional techniques based on accurately machined calibration objects. Thus, even in situations where precision equipment is not available, the proposed system would allow high-accuracy reconstruction.

It would also be desirable to combine multiple scans of the object corresponding to different object orientations or turntable heights. In this case, the ambiguities have the potential to be reduced once 3D correspondences between the scans are available.

The second novelty is the combination of shape constraints from both the silhouette and the surface texture. This allows the reconstruction of concavities in the object, while maintaining the strong consistency imposed by the silhouette. In contrast, state-of-the-art space carving algorithms — as represented by the papers of Kutulakos and Seitz (1998) and Broadhurst et al. (2001), for example — will often carve holes through the entire object if a single surface voxel fails its consistency test. The failure of the photo-consistency tests in these cases also represents an area in which our system would benefit from improvement. The current strategy for dealing with non-Lambertian surfaces and non-constant illumination is simply to assume that a locally affine intensity change is sufficiently general to account for all non-modelled effects. Where this model is too restrictive, it has been suggested[22] that a more general 'shuffle' model could be fitted. However, this generality necessitates a significant increase in the number of views required for

a stable solution, because it is applied independently at every voxel. A topic for future investigation is to impose a spatial prior on the parameters of such general models in order to stabilise these solutions. Analogously, spatial curvature priors may usefully be applied in many common cases. We continue to work towards an efficient solution to these open problems.

Note

We are grateful for permission to use the dinosaur sequence supplied by the University of Hannover, and for financial support from EU ACTS Project Vanguard, and the Royal Society. Special thanks are due to G. Parker and D. Kurtz of the Beazley Archive at the Ashmolean Museum, Oxford.

References

ARMSTRONG, M., ZISSERMAN, A. and HARTLEY, R. (1996), 'Self-calibration from image triplets', in B. Buxton and R. Cipolla (eds.), *Computer Vision: ECCV '96. 4th European Conference on Computer Vision, Cambridge, UK*, Lecture Notes in Computer Science 1064–5: 3–16.

BAUMGART, B. G. (1974), 'Geometric modelling for computer vision', PhD thesis (Stanford University, Palo Alto).

BEARDSLEY, P. and ZISSERMAN, A. (1995), 'Affine calibration of mobile vehicles', in R. Mohr and W. Chengke (eds.), *Europe-China Workshop on Geometrical Modelling and Invariants for Computer Vision* (Xi'an, China): 214–21.

BEARDSLEY, P., TORR, P. and ZISSERMAN, A. (1996), '3D model acquisition from extended image sequences', in B. Buxton and R. Cipolla (eds.), *Computer Vision: ECCV '96. 4th European Conference on Computer Vision, Cambridge, UK*, Lecture Notes in Computer Science 1064–5: 683–95.

BOYER, E. (1996), 'Object models from contour sequences', in B. Buxton and R. Cipolla (eds.), *Computer Vision: ECCV '96. 4th European Conference on Computer Vision, Cambridge, UK*, Lecture Notes in Computer Science 1064–5: 109–18.

BROADHURST, A., DRUMMOND, T. and CIPOLLA, R. (2001), 'A probabilistic framework for space carving', *Eighth International Conference On Computer Vision (ICCV-01), July 7–14, 2001, Vancouver, British Columbia, Canada*, IEEE Computer Society, vol. 1: 388–93.

CROSS, G. and ZISSERMAN, A. (2000), 'Surface reconstruction from multiple views using apparent contours and surface texture', in A. Leonardis, F. Solina and R. Bajcsy (eds.), *Confluence of Computer Vision and Computer Graphics* [proceedings of the NATO Advanced Research Workshop on Confluence of Computer Vision and Computer Graphics, Ljubljana, Republic of Slovenia, 29–31 August 1999] (Dordrecht and London): 25–47.

---

[22] Kutulakos and Seitz (1998).

DEMEY, S., ZISSERMAN, A. and BEARDSLEY, P. (1992), 'Affine and projective structure from motion', in D. Hogg and R. Boyle (eds.), *BMVC92: Proceedings of the British Machine Vision Conference...1992...Leeds* (London): 49–58.

FAUGERAS, O. D. (1992), 'What can be seen in three dimensions with an uncalibrated stereo rig?', in G. Sandini (ed.), *Computer Vision: ECCV '92. 2nd European Conference on Computer Vision: Papers*, Lecture Notes in Computer Science 588: 563–78.

FAUGERAS, O. D. and KERIVEN, R. (1998), 'Complete dense stereovision using level set methods', in H. Burkhardt and B. Neumann (eds.), *Computer Vision: ECCV '98. 5th European Conference on Computer Vision, Freiburg, Germany, June 1998*, Lecture Notes in Computer Science 1406 (Berlin and London): 379–93.

FITZGIBBON, A. W. and ZISSERMAN, A. (1998), 'Automatic camera recovery for closed or open image sequences', in G. Sandini (ed.), *Computer Vision: ECCV '92. 2nd European Conference on Computer Vision: Papers*, Lecture Notes in Computer Science 588: 311–26.

FITZGIBBON, A. W., CROSS, G. and ZISSERMAN, A. (1998), 'Automatic 3D model construction for turntable sequences', in R. Koch and L. Van Gool (eds.), *3D Structure from Multiple Images of Large-Scale Environments*, Lecture Notes in Computer Science 1506: 155–70.

GOLUB, G. H. and VAN LOAN, C. F. (1989), *Matrix Computations*, 2nd edn (Baltimore, MD).

HARRIS, C. J. (1990), 'Structure-from-motion under orthographic projection', in O. Faugeras (ed.), *Computer Vision: ECCV 90. 1st European Conference. Papers*, Lecture Notes in Computer Science 427 (New York and Berlin): 118–23.

HARTLEY, R. I. (1994), 'Euclidean reconstruction from uncalibrated views', in J. Mundy, A. Zisserman and D. Forsyth (eds.), *Applications of Invariance in Computer Vision*, Lecture Notes in Computer Science 825: 237–56.

HARTLEY, R. I. and ZISSERMAN, A. (2000), *Multiple View Geometry in Computer Vision* (Cambridge).

HARTLEY, R. I., GUPTA, R. and CHANG, T. (1992), 'Stereo from uncalibrated cameras', in *Computer Vision and Pattern Recognition: Conference. Papers*, IEEE Computer Society.

JIANG, G., TSUI, H., QUAN, L. and LIU, S. (2001), 'Recovering the geometry of single axis motions by conic fitting', in *2001 IEEE Conference on Computer Vision and Pattern Recognition, Kauai, Hawaii*.

KANG, S. B. (1997), *Quasi-Euclidean Recovery from Unknown but Complete Orbital Motion*, Technical Report 97–10 (DEC Cambridge Research Laboratory, Cambridge, MA).

KOENDERINK, J. J. and VAN DOORN, A. J. (1991), 'Affine structure from motion', *Journal of the Optical Society of America A* 8: 377–85.

KUTULAKOS, K. N. and SEITZ, S. M. (1998), *A Theory of Shape by Space Carving*, Technical Report CSTR 692 (University of Rochester).

LAURENTINI, A. (1994), 'The visual hull concept for silhouette-based image understanding', *IEEE Transactions on Pattern Analysis and Machine Intelligence* 16: 150–62.

MENDONÇA, P., WONG, K. and CIPOLLA, R. (2001), 'Epipolar geometry from profiles under circular motion', *IEEE Transactions on Pattern Analysis and Machine Intelligence* 23: 604–16.

MOONS, T., VAN GOOL, L., VAN DIEST, M. and OOSTERLINCK, A. (1994), 'Affine structure from perspective image pairs obtained by a translating camera', in J. L. Mundy, A. Zisserman and D. Forsyth (eds.), *Applications of Invariance in Computer Vision*: 297–316.

MUNDY, J. and ZISSERMAN, A. (1992), *Geometric Invariance in Computer Vision* (Cambridge, MA).

NIEM, W. and BUSCHMANN, R. (1995), 'Automatic modelling of 3D natural objects from multiple view', in Y. Paker and S. Wilbur (eds.), *Image Processing for Broadcast and Video Production: Proceedings of the European Workshop on Combined Real and Synthetic Image Processing for Broadcast and Video Production, Hamburg, 23–24 November 1994* (London).

POELMAN, C. and KANADE, T. (1994), 'A paraperspective factorization method for shape and motion recovery', in J.-O. Eklundh (ed.), *Computer Vision: ECCV '94. 3rd European Conference. Papers*, vol. 2, Lecture Notes in Computer Science 801: 97–108.

POLLEFEYS, M., KOCH, R. and VAN GOOL, L. (1998), 'Self calibration and metric reconstruction in spite of varying and unknown internal camera parameters', in *Sixth International Conference on Computer Vision...January 4–7, 1998, Bombay, India* (New Delhi and London): 90–96.

QUAN, L. and MOHR, R. (1992), 'Affine shape representation from motion through reference points', *Journal of Mathematical Imaging and Vision* 1: 145–51.

SAWHNEY, H. S., OLIENSIS, J. and HANSON, A. R. (1990), 'Description and reconstruction from image trajectories of rotational motion', in *Computer Vision: 3rd International Conference [Osaka]. Papers* (Los Alamitos, CA): 494–8.

SEITZ, S. M. and DYER, C. R. (1997), 'Photorealistic scene reconstruction by voxel coloring', in *1997 IEEE Computer Society Conference on Computer Vision and Pattern Recognition [Puerto Rico]*: 1067–73.

SHAPIRO, L. S., ZISSERMAN, A. and BRADY, M. (1995), '3D motion recovery via affine epipolar geometry', *International Journal of Computer Vision* 16: 147–82.

SHUM, H. and HE, L. (1999), 'Rendering with concentric mosaics', *Computer Graphics* 33: 299–306.

SULLIVAN, S. and PONCE, J. (1998), 'Automatic model construction, pose estimation, and object recognition from photographs using triangular splines', in *Sixth International Conference on Computer Vision...January 4–7, 1998, Bombay, India* (New Delhi and London).

SZELISKI, R. (1991), 'Shape from rotation', in *Computer Vision and Pattern Recognition [Lahaina, HI]*: 625–30.

SZELISKI, R. and KANG, S. B. (1993), *Recovering 3D Shape and Motion from Image Streams Using Non-Linear*

Least Squares, Technical Report 93/3 (DEC Cambridge Research Laboratory, Cambridge, MA).

TOMASI, C. and KANADE, T. (1992), 'Shape and motion from image streams under orthography: a factorization approach', *International Journal of Computer Vision* 9: 137–54.

TORR, P. H. S., FITZGIBBON, A. W. and ZISSERMAN, A. (1999), 'The problem of degeneracy in structure and motion recovery from uncalibrated image sequences', *International Journal of Computer Vision* 32: 27–44.

TORR, P. H. S. and MURRAY, D. W. (1997), 'The development and comparison of robust methods for estimating the fundamental matrix', *International Journal of Computer Vision* 24: 271–300.

VIÉVILLE, T. and LINGRAND, D. (1995), *Using Singular Displacements for Uncalibrated Monocular Vision Systems*, Technical Report 2678 (INRIA).

ZISSERMAN, A., BEARDSLEY, P. and REID, I. (1995), 'Metric calibration of a stereo rig', in *IEEE Workshop on Representation of Visual Scenes, [Boston]* (Los Alamitos, CA): 93–100.

ZISSERMAN, A., LIEBOWITZ, D. and ARMSTRONG, M. (1998), 'Resolving ambiguities in auto-calibration', *Philosophical Transactions of the Royal Society of London A* 356: 1193–1211.

CHAPTER 7

At the Foot of Pompey's Statue: Reconceiving Rome's *Theatrum Lapideum*

Hugh Denard[1]

In the course of his career, Rome awarded Gnaeus Pompeius Magnus ('the Great') three triumphs for his military conquests in Africa, Europe, and Asia. These extraordinary displays were designed to enhance Pompey's standing and influence in Rome by impressing upon spectators the magnitude and importance of his achievements. But Pompey quickly discovered that the political popularity these spectacular displays could procure was as ephemeral as the triumphal processions themselves. On 29 September 55 BC, therefore, six years after his third triumph, Pompey invited the citizens of Rome to celebrate a festival marking the dedication of a vast new amenity designed by him for their pleasure and benefit.

The city had never seen anything like it. Pompey's new precinct in the Campus Martius contained the first permanent theatre in Rome. Unsupported by natural hillside, this gigantic structure housed a stage some 290 feet wide, and may have accommodated up to 40,000 spectators.[2] Crowning the auditorium, a temple dedicated to Pompey's patron Venus Victrix surveyed the proceedings. Second in height only to the Capitol, it was accompanied by subsidiary temples to Honos, Virtus, and Felicitas.[3] According to Valerius Maximus, Pompey caused water to run through the *cavea* to refresh the audience.[4] These elements themselves would have been sufficient to inscribe indelibly the magnificence of Pompey's achievements upon the memories and imaginations of his own and subsequent generations, but behind the stage building Pompey also created a great colonnaded public park—the first such in Rome. The elegantly landscaped garden, planted with exotic trees Pompey had brought back from Asia and Africa, and cooled by fountains and running water, was ornamented by trophies signifying Pompey's conquests, and a thematically linked ensemble of sculpture. For writers such as Ovid, Propertius, and Martial, it was noted as a preferred venue for amorous assignations and companionable walks in the shade.[5] During intervals in theatrical festivities, Vitruvius tells us, audiences could withdraw into this park without crowding into the surrounding city streets. The park also acted as a capacious backstage area for performers, scenery, and props which on more spectacular occasions could number thousands. The dedicatory games themselves included a performance of Accius' *Clytaemnestra* in which Agamemnon's triumphal train boasted no fewer than 600 mules, according to Cicero, who also describes a performance of *The Trojan Horse* at the same festival in which three thousand bowls were displayed on stage.[6] The north side of the park was bounded by a Hecatostylum with golden tapestries from Pergamon suspended between its hundred columns, providing a regal backdrop to Pompey's Mouseion. Finally, at the far end of the park, Pompey erected a Senate House. By positioning the entire precinct outside the city boundary (*pomerium*), Pompey ensured that he could attend the theatre and meetings of the Senate without surrendering his power of command (*imperium*) as General.[7]

This procession of artistic, botanical, and architectural treasures was ideologically encoded in ways that the visually sophisticated Romans were well accustomed to reading, and provided a potent aesthetic and political response to the transience of the triumphal procession.[8] In gratitude, the city set up a statue of Pompey in his Senate House; it was at the foot of this statue, eleven years later, that Pompey's conqueror, Julius Caesar, was himself slaughtered. The Theatrum Pompeianum was for centuries the site of many of the most important events in the cultural and political life of the city of Rome (Nero himself flouted all custom by performing upon its stage, an act calculated to provoke the elites and delight the masses),[9] and as the first major example of Roman imperial architecture, it decisively influenced the style of

---

[1] My grandfather Francis William Rogers Brambell died at the height of his powers in 1970, the year I was born. Since he was a Fellow and Medallist of the Royal Society, this seems an appropriate occasion to remember him; I like to imagine this paper as a sort of recompense, however formal and fleeting, for conversations we might have had.

[2] Pliny (*NH* 36.115) says that the theatre could seat 40,000 spectators. While this figure has long been doubted, it now appears that the *cavea* may indeed have accommodated these numbers, although a lower capacity—in the range of 20,000–30,000—cannot as yet be ruled out.

[3] The temple to Venus was probably incomplete at the time of the games. Gell. (*Noctes Atticae* 10.1.6) records that it was eventually dedicated on 12 August 52 BC. There may have been further temples *summa cavea*: see Steinby (1999).

[4] Val. Max., 2.4.6.

[5] Ovid, *Ars Amatoria* 3.387; Prop., *Elegiae* 2.32.11–20; 4.8.75; Mart., *Epigrammaton* 11.14.10.

[6] Cic., *Ad Fam.* VII.1.

[7] For a fuller account, see Beacham (1999a), chapter 2.

[8] See Gleason (1994), 13–27.

[9] Dio, 62.29.1; Suet., *Vit.* 4; Tac., *Ann.* 16.4.

Figure 7.1 Aerial photograph of the area of present-day Rome corresponding to the Temple of Venus Victrix, *cavea*, orchestra, and *scaenae frons* of the Theatre of Pompey. Copyright University of Warwick.

Rome's urban development. When Vitruvius wrote his treatise, *De Architectura*, it is likely that Pompey's edifice — which he mentions by name — formed the basis for his account of theatres. Through Vitruvius, it arguably became the prime architectural model for vast numbers of theatres constructed throughout the empire. As late as its final restoration in the sixth century AD under the Ostrogoths, the theatre was still sufficiently imposing for Theodoric's chancellor, Cassiodorus, to describe: 'caves vaulted with hanging stones, so cleverly joined into beautiful shapes that they resemble more the grottoes of a huge mountain than anything wrought by human hand'.[10]

It is astonishing to learn, therefore, that despite its great historical, architectural, and cultural importance,[11] until now there has been no modern scientific survey of the remains of the theatre. Most recent discussions have been based upon the limited excavations and site plans of Victoire Baltard of the École des Beaux Arts, working in the first decades of the nineteenth century, or the subsequent, and partly derivative, work of Luigi Canina. But no comprehensive analysis of the site has been conducted, and consequently questions of major importance remain unsolved. The situation is not improved by the fact that there is now little to see above ground: over the centuries, the remains of the theatre were gradually subsumed into the medieval and Renaissance structures which occupy the site. Despite plans drawn up in the Fascist period which would have stripped away these post-antique elements, the theatre has thus far obstinately refused to relinquish its acquired architectural clothing (Figure 7.1). The monument cannot now

be extensively excavated. Moreover, as this area of Rome becomes increasingly affluent, occasionally interesting details previously obscured or lost are uncovered, but more frequently vital but less glamorous archaeological remains of the theatre become the victims of development. In the relatively short period of our own work, we have seen *opus reticulatum* walls belonging to the theatre disappear beneath new plastering, or sections of the theatre structure become obscured by modern 'improvements'. Contemporary analysis of the site is therefore framed by an ever-diminishing window of opportunity.

Virtual reality (VR) technologies, however, have greatly enhanced our capacity to study and to understand such structures. In 1998, with funding from the British Academy and the Leverhulme Trust, Professor Richard Beacham of the University of Warwick conducted a preliminary study to ascertain the existing state of knowledge on the theatre. In the spring of 1999 the Arts and Humanities Research Board granted him substantial funds to co-ordinate, together with Professor James Packer (Northwestern University), a major interdisciplinary study of the monument that would be aided by three-dimensional computer modelling. The Pompey Project both benefits from and contributes to a wider programme of digital research being conducted at the University of Warwick in which the application of IT — particularly VR — to humanities research is being explored.[12] Working on the project is an international interdisciplinary, scholarly team including archaeologists, theatre historians, art historians, urban historians, and a number of 3D modellers who, closely collaborating with the project archaeologists and architects, have acquired considerable expertise in working with antique materials. In the summer of 2002, funded by various sources in the USA, Professor Packer directed the first excavation on the site in 150 years,

---

[10] Cassiodorus, *Variae* 4.51.

[11] Ann Kuttner, in analysing the textual history of the Theatre of Pompey since its construction in 55 BC, finds 'what may be the broadest and most varied textual dossier on any *monumentum* at Rome, accreted in a range of genres over the next half-millennium' (Kuttner, 1999, 344).

[12] For a recent survey of work, see Beacham (1999b).

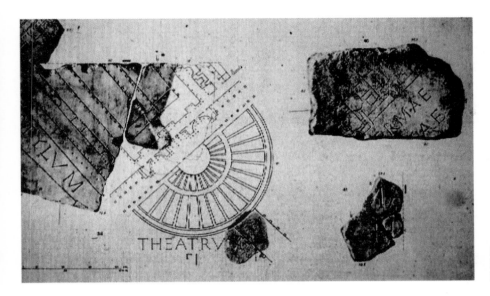

Figure 7.2 Fragments of the Marble Plan: Museo della Civiltà Romana, Rome. See Almeida (1980).

the VR-enhanced research process having enabled the archaeological experts to pinpoint the optimum location. This has significantly contributed to our understanding of parts of the structure which hitherto have been obscure, as well as illuminating aspects of the post-antique history of the site. Permission from the authorities in Rome has been obtained for a further four seasons. When this work is complete, the project will produce — in addition to a scholarly monograph — a highly sophisticated and integrated electronic resource spanning the entire history of the site, from antiquity to the present. It will include 3D computer models, digital modelling of the structure's acoustics, images of artefacts, a database of all published references to finds of architectural elements likely to have been contained in the complex, all known previous textual references to and studies of the site, a comparative history of scholarship on the site undertaken in part by creating 3D models of previous attempts to reconstruct the theatre graphically, and a comparative study — also aided by 3D modelling — of the theatre-architectural antecedents to, and descendants of, the Theatre of Pompey.

Virtual reality technologies have been bringing about a profound revolution in the ways in which knowledge is produced. Virtual reality can enable the formation of new knowledge. By making knowledge visible, for example by translating archaeological survey data into three-dimensional form, it offers new ways of knowing; and by making visible the unknown, for example by enabling researchers to test hypotheses of possible reconstructions of lost or hidden areas of structures in three dimensions, it promises to make knowable things that hitherto were unknowable. In my presentation at the joint British Academy–Royal Society symposium, I offered a detailed account of the research process through which certain parts of the theatre have been digitally modelled in 3D. I am able to reproduce here only a small sample of the collection of images and computer-rendered animations upon which this narrative is dependent, so this section of this paper will be brief. Thereafter, I offer a discussion, developed from the concerns raised in the paper, of

some of the methodological and theoretical issues which we have encountered and explored in the course of the project.[13]

Archaeology and Virtual Reality

The archaeological work of the project began with a programme of research into the Theatre of Pompey's *scaenae frons* (stage building) as rebuilt after the fire of AD 80. Fortunately, this stage building is one of the structures for which we have substantial actual and recorded fragments of the Severan Marble Plan — a detailed and to-scale plan of the city of Rome carved onto stone slabs between AD 203 and 211 (Figure 7.2). Whatever new data may emerge from the recent three-dimensional scan of the plan by a team from Stanford University,[14] it already clearly shows the layout of the *scaenae frons* and part of the *porticus post scaenam* of the theatre.

Drawing upon the Marble Plan, together with other surviving archaeological evidence, reinforced by the documentation of nineteenth-century investigations, Professor Packer and John Burge, who is an expert in creating VR reconstructions of Roman architecture, have been able to calculate the size and dimensions of the theatre with considerable precision. According to the plan, the rear façade of the stage building is almost identical in layout and size to the façade of the Basilica Ulpia in the Forum of Trajan, dedicated just a few decades later, in AD 112. Archaeological finds on both sites confirm the correspondence. From this comparison, Packer and Burge were able to reconstruct plausibly the appearance of the rear of the Theatre of Pompey's *scaenae frons* (Figure 7.3), which would have

---

[13] The images here represent work in progress, and show only the current research hypothesis pending the completion of the project's archaeological study.

[14] For an account of the Stanford digitisation of the Marble Plan, see: http://graphics.stanford.edu/projects/forma-urbis/.

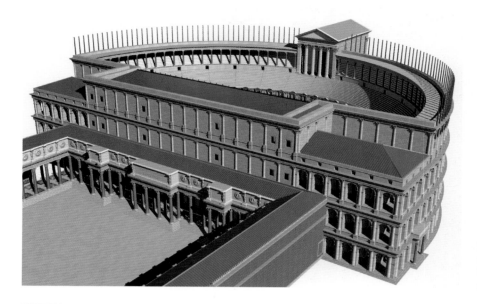

Figure 7.3 Theatre of Pompey: *porticus post-scaenam* and Temple of Venus Victrix. 3D model by John Burge. Copyright University of Warwick.

Figure 7.4 Theatre of Pompey: orchestra and *scaenae frons*. 3D model by John Burge. Copyright University of Warwick.

visually dominated the park area. Following this hypothesis they could extrapolate the proportions of the *scaenae frons* itself, together with its precise configuration of columns and *exedrae*. The calculation of the elevation of the *scaenae frons* involved comparing proportions given by Vitruvius to those of surviving Roman *scaenarum frontes*, and then reconciling these calculations with known fragments of column shafts, capitals, cornices from the Theatre of Pompey, and an architrave found near the original location of the theatre's stage building. Ultimately, by combining this evidence, Packer and Burge have been able to propose a persuasive reconstruction of the layout, proportions, and distribution of materials (*giallo antico* for the first order, *cipollino* for the second, *porphyry* for the third, and *africano* for the door order) within the elevation. In John Burge's VR model, textures simulating these stone types are added to the virtual structure in order to represent this hypothesis in visual form (Figure 7.4).

If the reconstruction of the *scaenae frons* demonstrates the way in which an apparently very small amount of direct archaeological evidence can be made to yield great returns, in the case of the *cavea* the evidence is not so thin on the ground. Diners in the Ristorante Pancrazio and patrons of various shops, cafes, and galleries in largo del Pallaro, via di Grotta Pinta, and via dei Chiavari will be familiar with the way in which later structures have encrusted themselves around the theatre's ruins. Here, then, the project's first task was to establish the precise architectural relationship between the ancient theatre and later structures. This necessitated an aerial survey of their roofs, which we conducted by helicopter (Figure 7.1). During our interior survey of the post-antique buildings, we discovered that some remains of the theatre extend upwards five storeys through the Palazzo Pio, the Renaissance building positioned over the area formerly occupied by part of the *cavea* and Temple of Venus Victrix. Working closely

Figure 7.5 Existing state of segment of *cavea* substructure, Hotel Teatro di Pompeo. Copyright University of Warwick.

Figure 7.6 Theatre of Pompey: exterior of *cavea*. 3D model by John Burge. Copyright University of Warwick.

with James Packer, project architect and archaeological historian Dario Silenzi then conducted a full survey of every section of the *cavea* to which we could gain access (Figure 7.5). This resulted in scaled sections of every ancient wall on the site, establishing their precise character and state of preservation. Silenzi also prepared a full digital photographic record of the interiors of each ancient chamber. This, in turn, has enabled John Burge to create a three-dimensional model of the remains of the theatre's *cavea*, providing an unrivalled resource for the analysis of the monument's existing state, and simultaneously enabling the collaborators to develop hypothetical reconstructions of the unknown, using precise visual data.

In one of the basement rooms accessible through Pancrazio's restaurant, the top of an ancient door is exposed. By combining virtual reconstructions of the known existing state of the monument with hypotheses of what may lie beyond, Packer, Silenzi, and Burge identified

this as the optimum site for the new archaeological excavation. It is hoped that the excavation will enable us to ascertain a number of vital facts about the *cavea* for which the evidence, notwithstanding the extent of the remains, is at present unclear. It might even be possible to expose two bays of the theatre's actual exterior façade, which we have never seen (hypothetically reconstructed in Figure 7.6). It will assist in determining the stratigraphy of the theatre and porticus, and that of the entire building complex, from the Senate House (now beneath part of largo Argentina) to the temple (campo dei Fiori). More locally, it should reveal new information about the ground-floor plan of the structure under the *cavea*, with regard to stairways and internal corridors. The significance of this lies at least partly in the fact that this theatre was probably the first example of a system of audience circulation that was widely imitated later, and continues to be used in the present day in modern sports stadiums.

Information Technology Enhanced Research

When we began our work we tended to view information technology primarily as a means of enhancing essentially traditional research methods. Our thinking and methods have undergone considerable evolution since then, giving rise to some new perceptions about the nature of VR-based knowledge, and a keen sense of how in this field of work traditional boundaries, between data and interpretation, evidence and argument, researcher and technician, are undergoing rapid and profound transformation.

Firstly, the technology both enables and requires the project to be inherently interdisciplinary. For us, that has meant the creation of a large research team, linking VR modellers to archaeologists, database experts to theatre historians, archaeological surveyors to urban historians, all joined by the shared need to produce a 'virtual' structure. For the collaborators, the need to understand and respond co-operatively to the working methods of such a range of colleagues has been intellectually and imaginatively stimulating, opening up new modes of perception and ways of thinking. Only a few years ago, these specialists would have had little opportunity or reason even to discuss their work with one another, much less engage in a process of intensely creative collaboration.

IT has also enhanced our ability to process and manipulate huge datasets of several information-types in 3D, leading to better analysis and hypotheses. Three-dimensional models share certain of the properties, demands, and advantages of CAD drawings: both rely on precise sets of co-ordinates, and require an absolute degree of exactitude — they are unforgiving in this respect. Consequently the data used to inform such models must be vigorously evaluated and co-ordinated. In addition, because 3D models require the spatial relationship between objects to be calculated in three dimensions, problems of relation, proportion, measurement, and design which are difficult to identify during the creation of 2D representations become immediately and persistently apparent. VR thus precipitates and enables a constant re-examination and reinterpretation of data in ways previously difficult or labour-intensive to the point of impossibility. Unlike manual drawings or solid models, virtual models can easily and quickly be altered to incorporate new data. The consequences of such modifications for other elements in the model can instantly be seen and assessed, enabling rival hypotheses to be evaluated quickly and accurately.

Furthermore, 3D modelling allows different forms of model to be produced according to different modes of enquiry: e.g. CAD drawings for accurate representation of the volume and measurements of a building; cut-away models to enable the user to investigate architectonic data and hypotheses; or other types of models to simulate lighting or acoustics or to represent changes to a given space over time. Massing models of other cognate sites allows us to analyse possible architectural antecedents and descendants of the building, and thus assist in delineating its architectural genealogy. And as we have seen, 3D modelling can also enable the project archaeologists to determine more precisely the optimum locations for new excavations, and to assess their probable impact and value.

The project also exemplifies the considerable benefits of digital dissemination of scholarly research, including the capacity to publish all of the data produced by and for the project economically and effectively, and to include in the publication a wide range of media types. Through including comprehensive databases, both free-standing and linked to models, publication in this form combines and significantly exceeds the strengths of a scholarly monograph, excavation notes, documentation, and photographic record. The technology also greatly enhances the means available to researchers for querying the research data, enabling more varied and sophisticated uses of the published record, and the enormously useful capacity to zoom in almost indefinitely on details in 3D models. For educational or display purposes, the content-dense, interactive moving images produced by VR technologies are vastly more informative and engaging than still images, particularly when digital audio and lighting technologies can be brought to bear upon them. The project documentation presented on the World Wide Web can be updated in response to advances in scholarship and the contributions of other researchers, and perhaps foster the development of an online community of criticism and debate on the theatre; we hope that the website may become a prime, dynamic *locus* of scholarship, transcending traditional conceptualisations of the relationship between research and publication.

The Trouble With Textuality...

Despite all these advantages, however, there are dangers. This new kind of digital 'text' can be a source as much of anxiety as of information. In bringing together both the information structures of the original building and a simulation of its decorative elements, 3D models and their derivatives (rendered images, animations, etc.) acquire a seductive 'persuasiveness' that can easily render invisible crucial distinctions between known fact, scholarly deduction, and creative — albeit educated — guesswork, and erode a sense of their provisional nature as research hypotheses. As suggestive indices to a possible architectural past they function quite well, but unless they can in some way display to their users the state of knowledge that they *truly* represent, their value as instruments of scholarly communication — much less enquiry — is ultimately dubious. In the case of the Pompey Project this 'persuasiveness' is particularly acute, as a result of, on the one hand, the extraordinarily life-like detail of the models created by John Burge[15] and,

---

[15] What the rendered images reproduced here cannot show is that every contour of every capital and frieze is fully modelled in three dimensions (a single Corinthian capital currently occupies some 50 megabytes of disk space). John Burge used Silicon Graphics Octane computers with dual Pentium 3 processors running at

Figure 7.7 Preliminary massing model of the Theatre of Pompey superimposed upon an aerial photograph of the modern city. Copyright University of Warwick.

on the other, the immersive, real-time experience of the theatre that the viewer can obtain from the VRML model (Figure 7.7) prepared by Drew Baker at the University of Warwick.[16]

Nor is it difficult to see how the task of translating archaeological survey data into a form which boasts such a powerful capacity to simulate reality makes the lure of a positivist paradigm of reconstruction all too attractive for the project collaborators. Such an approach, which conflates 'knowledge' with 'truth', could occlude both the project's methodological and interpretative biases and the provisional and partial nature of the knowledge that it produces. The structure of any interface or database produced by the project would naturally reflect such problematic values, to be recognised and resisted only by the most critically self-conscious of readers (and even then belatedly).

To some extent, then, the newness of VR-based research places us in the anomalous position of being authors of a new type of 'text' that we have not as yet fully learnt how to read. For this reason, as our work has progressed we have had to develop new ways of conceptualising our research in response to a growing need to understand the nature and agency of the digital 'texts' we are producing.[17] The perceived characteristic qualities of any text — its 'textuality' — can largely be defined in terms of genre and medium; these two types

of structure strongly condition the parameters within which texts are produced and received, and therefore the types of knowledge that 'creators' and 'readers' will produce in relation to them. One of the most complex phenomena that we have encountered is the degree to which different forms of textuality, whether real or virtual, enable (indeed, insist upon) correspondingly different ways of producing knowledge — different ways to *epistemes*. Since we are engaged in the creation of relatively new forms of texts (e.g. three-dimensional digital objects, rendered images, and animations) and in placing them within further framing texts (e.g. user interfaces, hyperlinked structures) with a view to generating and disseminating new knowledge and new types of knowledge — i.e. with an explicit, epistemological aim — it is incumbent upon us to seek to appreciate the distinctive textual characteristics of these VR texts and their implications.

Alongside our eagerness to tap the extraordinary possibilities for the modelling of knowledge that the virtual realm offers, therefore, has been an equally pressing concern to explore how digital technologies can be harnessed to keep visible both the limitations and the positionality of knowledge. What follow are a number of ideas and strategies that we have developed in response to these concerns.

Firstly, for authors and readers of the Pompey Project, the intensively interdisciplinary nature of the project implies a heterogeneity of critical perspectives which in itself visibly militates against the formation of a perceived methodological or interpretative orthodoxy, and thus serves to undermine any imagined claim of a single text within the resource — whether literary or graphic — to the status of 'definitive text'.

Secondly, while our work to date has necessarily concentrated on producing models of the main research hypotheses in order to facilitate our own archaeological research, we have increasingly generated models which represent multiple hypotheses in regard to sections or

---

850 MHz, assisted by 2 gigabytes of RAM, but even at these specifications it took about an hour to render one of these images at a mere 72 dpi. It will be some time, one supposes, before the average desktop computer will be able to navigate these colossal models in real time.

16 Drew Baker's model, created using virtual reality modelling language (VRML), occupies a mere 119 K (or 20 K compressed), and allows readers to walk or fly in real time to any position in the building. This immersive experience offers a completely different kind of 'persuasiveness' to that of the detailed Burge models, but one which is no less problematic with regard to the questions raised here.

17 For example, Denard (2002).

aspects of the complex for which the archaeological evidence is insufficient to reach firm conclusions, or where *comparanda* suggest a number of equally plausible options. Further models may indicate different levels of archaeological probability within the digital reconstructions. The incorporation of alternative perspectives such as these at a high level within the information hierarchy of the final publication will further combat any perception that the main hypothesis makes any claims to being 'definitive'. In time we hope that these models will respond to post-publication feedback from users and experts, leading to digital realisation of alternative interpretations of the data; they may eventually even permit users to apply different textures and patterns — perhaps even proportions — according to their preferred interpretation of the available data.

Thirdly, we constantly assert the provisional character of our hypotheses by locating them within a history of scholarship on Pompey's theatre. To this end, we have created 3D models of all significant previous attempts to reconstruct the theatre graphically, and have digitised a considerable collection of scholarship and documentation on the site which we hope to publish as part of our dataset. While such a teleological narrative might at this proximity seem to be a strategy designed to aggrandise our work as the final culmination of a tradition of scholarship, we trust that scholarly and technological developments will quite quickly enable our work to be read in a longer perspective: namely, as the most recent, detailed, and comprehensive study of this site to date, and a considerable resource for future research, but also necessarily, and ineluctably, provisional.

Finally, in order to demonstrate that every on-screen image is neither more nor less than an informed and closely argued interpretation and/or hypothesis, we are incorporating into the digital publication comprehensive documentation setting out the investigative, methodological, and interpretative processes that have led to the creation of each element of each model. By embedding these data dynamically into the user interface and by integrating them in various ways into the three-dimensional objects or rendered images, our database and interface technologies can assist us in asserting the interrogative, analytical, and interpretative nature of the project's work. Moreover, by careful configuration of the structures of the digital publication — in particular by enabling expert users to explore the data effectively, but unencumbered by highly prescriptive threading structures — we can go a long way towards liberating the reader of the resource from the critical perspectives and agendas of its authors; such readers will be able to interpret and exploit the data according to their own needs, agenda, and contexts.

Conclusion

Our engagement with virtual reality has impacted upon every conceivable aspect of the project's work. It has demonstrably enhanced the research process in both efficiency and efficacy, and will certainly enhance the dissemination process. It may, perhaps, contribute to the creation of a more open conceptualisation of publication as feedback from users is incorporated, and as the models migrate from generation to generation. VR technology has also been a hard taskmaster, requiring of the collaborators exacting co-ordination of technical specifications across a diverse group of disciplinary practices, and exhaustive strategic planning and communication to ensure that the dictates and implications of virtual-reality-orientated research are fully recognised and taken into consideration by each of the partners.

As more and more scholarship either takes place in or results in virtual reality, we must face the challenge of developing ways of both creating and reading such texts with a keen attentiveness to the complexity of their unique textuality. Our determination is that the persuasive quality of these images should be harnessed to provide not just a compelling aesthetic experience for the viewer, but also a persuasive scholarly 'text', conjoined to an extraordinarily rich collection of resources, so that future generations of scholars will continue to benefit from what we have begun.

References

ALMEIDA, E. (1980), *Forma Urbis Marmorea: aggiornamento generale* (Rome).

BEACHAM, R. (1999a), *Spectacle Entertainments of Early Imperial Rome* (New Haven and London).

BEACHAM, R. (1999b), ' "Eke out our performance with your mind": reconstructing the theatrical past with the aid of computer simulation', in T. Coppock (ed.), *Information Technology and Scholarship: Applications in the Humanities and Social Sciences* (Oxford): 131–54.

DENARD, H. (2002), 'Virtuality and performativity: recreating Rome's Theatre of Pompey', *Performing Arts Journal* 24/1.

GLEASON, K. L. (1994), '*Porticus Pompeiana*: a new perspective on the first public park of ancient Rome', *Journal of Garden History* 14/1: 13–27.

KUTTNER, A. (1999), 'Culture and history at Pompey's museum', *Transactions of the American Philological Association* 129: 343–73.

STEINBY, E. M., ed. (1999), *Lexicon Topographicum Urbis Romae* 5: 36 (Rome).

CHAPTER 8

Modelling Sagalassos: Creation of a 3D Archaeological Virtual Site

Luc Van Gool, Marc Pollefeys, Marc Proesmans & Alexey Zalesny

I. The Murale Project

This paper describes part of the planned contributions of Murale, an IST (Information Society Technologies) project funded by the European Commission, in order to advance the use of computer technology in archaeology.

The Murale consortium consists of the following partners: Brunel University (UK), ETH Zurich (Switzerland), Eyetronics (Belgium), Imagination (Austria), the Technical University of Vienna (Austria), the University of Graz (Austria), and the University of Leuven (Belgium). The main areas of expertise of its researchers are archaeology, computer vision, and computer graphics. Murale is about the development of technology, but from the start archaeologists also wished to focus on its practical application on a test site. This site is the ancient city of Sagalassos, in what is now the southern part of Turkey.

The site at Sagalassos is one of the largest archaeological projects in the Mediterranean, led by Professor Marc Waelkens of the University of Leuven. Sagalassos thrived during a period of about 1000 years (fourth century BC–seventh century AD), and during that period became one of the three greatest cities of ancient Pisidia. Sagalassos lies about 100 km to the north of Antalya, in the Turkish province of Burdur. Figure 8.1 shows this location in more detail. The ruins of the city lie on the southern flank of the Aglasun mountain ridge (western part of the Taurus mountains) at a height between 1400 and 1650 m. Figure 8.2 shows the valley with Sagalassos against the mountain flank. During its existence the city came under the military, political, and cultural influence of a series of foreign powers. In 333 BC it was defeated by Alexander the Great. Sagalassos was already an important city at that time. In the subsequent period it changed hands on several occasions between the successor kingdoms of Alexander's splintered Macedonian empire. From 189 BC onwards, the Romans directly intervened in the region until in 133 BC it finally became formally part of the Roman state. In 25 BC the emperor Augustus created the province of Galatia, which also incorporated Pisidia. It goes without saying that the city had changed substantially during this extended period, with its heyday around the second century AD. Murale will not be able to cover this whole evolution, and will focus on some selected periods.

The issues of how to record and visualise the finds of excavations are as old as archaeology itself. Several aspects of Murale will allow archaeologists to solve old tasks with new means, while some others add components that had been out of reach before. These are the goals that Murale plans to demonstrate at Sagalassos:

1 To provide tools that archaeologists themselves can use *in situ*. All devices should be easy to bring to the site, they should work robustly under conditions that can be quite adverse (sun heat, dust, moisture), and they should be easy to operate and carry around. Most excavation campaigns have little financial means and, hence, the devices should also be cheap.

2 To generate 3D models of objects at different scales: from landscapes, over buildings and statuary, to small finds such as potsherds. Acquisition should be fast enough in order not to interrupt the excavation longer than traditional methods would. The novel 3D technology should also introduce new possibilities, such as matching parts for virtual restoration/*anastylosis* and the creation of 3D stratigraphic records. The most obvious use of the 3D data is the realistic visualisation of the scene. This is useful for the public and the archaeologists alike. For example, archaeologists may use the terrain model to guess the positions of invisible infrastructure, such as long-gone roads, or they may use the 3D city models to understand why fortifications such as towers were built at specific places.

3 To provide a database for efficient storage and retrieval of finds. Such databases can be made available over the Internet. This turns them into powerful research tools, as archaeologists want to compare their finds against those found elsewhere. In the case of Murale the emphasis will be put on potsherds, as Sagalassos was a major production site of pottery. Other excavations can use the elaborated typology and corresponding time scale to date layers in which Sagalassos ceramic ware is found. In order to compare pot shapes special search tools will be developed.

4 To support different functions by integrating the 3D data and the database. First, by having chronological information with the data, visualisation for different time periods is made possible. When the user selects a certain time period, the system itself can look for buildings, objects, and so on from the specified period. Another way of linking the database

Figure 8.1 Sagalassos, the primary test site of the Murale project in southern Turkey.

Figure 8.2 Overview of the Sagalassos site.

with the 3D models is to provide annotations. If the user clicks on parts of the scene, additional information pops up.

In summary, Murale hopes not only to offer the public a more convincing and enticing impression of how Sagalassos developed over the centuries, but just as much to provide the archaeologists with tools that are effective and efficient in the field.

In this paper the emphasis is on work carried out by three of the partners: ETH Zurich, Eyetronics, and the University of Leuven. The results mainly pertain to 3D acquisition and visualisation, and to two complementary issues in particular:

- *3D shape acquisition*: model extraction from images (buildings, landscape) and a 3D camera (statues, sherds)
- *the surface texture*: analysing and synthesising textures to emulate building materials and landscapes.

These topics are described in Sections II and III, respectively. Only preliminary results can be shown at this point, as most of the work remains to be done.

Section IV concludes the paper and describes some future work.

II. 3D Shape Acquisition

This section describes the two major 3D acquisition methods of the Murale project. The first is a 'passive' one, the second is 'active'. The terminology 'active' *vs.* 'passive' in this context means 'with' *vs.* 'without' special illumination. Although passive techniques tend to be more flexible, the use of active techniques has some important advantages. The passive technique is used mainly for big structures, such as buildings and the terrain. The active method is used for smaller and more intricately shaped objects, such as potsherds, building ornaments, and statues. Both approaches are now described in more detail.

Passive Method: Shape from Multiple Photographs

The University of Leuven has developed a technique that allows us to create 3D textured models of the existing site, for example of the ruins or the terrain. It uses

(a)

(b)

Figure 8.3 (a) Three views of the Roman bathhouse at Sagalassos. (b) Three views of the model constructed from six images, three of which are shown in (a).

images as its only input. Hence, the archaeologists need simply take photographs of interesting finds, as they typically already do. With the purpose of 3D modelling, it may be necessary to take more, however, from a sufficient number of viewpoints. In general this can be done rather quickly and easily, although on-site experience has already shown that it may not be easy to cover a sufficient range of viewpoints once excavations are well under way. Walking around an excavation pit, for example, may prove difficult in the presence of a surrounding wall or neighbouring pit. Nevertheless, this method imposes a substantially lower overhead than traditional photogrammetric and computer vision approaches, as the motion and parameters of the camera need not be known. Motion parameters (rotation and translation) are typically called 'extrinsic' camera parameters. Camera-specific parameters such as pixel dimensions and focal length are referred to as 'intrinsic' parameters. The 3D modelling software extracts these parameters automatically, together with the 3D shape of the scene. As a result, existing footage can also be used to model scenes in 3D that no longer exist.

The first step is the automatic matching of corner-like structures between the different images, similar to the work by Armstrong et al.[1] The system will try to ensure that these features are well spread over the image, i.e. that they are not all selected in a small region where many obvious corners cluster. The correspondences between the corners are found by correlating the intensity patterns in small neighbourhoods around them. This only works if the motion between subsequent views is small. Therefore, additional methods are being developed to find correspondences between images taken from viewpoints far apart, using so-called 'invariant neighbourhoods' as an alternative for the corners. These are image patches that automatically change their shape with the viewpoint, in order to systematically cover the same physical patches on objects' surfaces.[2] Experiments have shown successful matches for differences in viewing angle up to 150°. Given the previous remark about potential difficulties in taking photographs

that are regularly spaced around the scene of interest, such wide-baseline algorithms can be expected to be particularly useful in applications like these.

Once the primary features have been matched, the system can put in place geometric constraints that facilitate the search for further correspondences. Epipolar constraints restrict the search for corresponding points to a line in the other image, and trifocal constraints lay links between triples of images. The system tries to generate dense correspondences, i.e. to find a correspondence for almost all pixels.[3] Points for which correspondences can be found can in the final step be reconstructed in 3D.

Even if one doesn't know the camera parameters, a pair of images already allows the system to produce a reconstruction that is correct up to some unknown projective deformation. Self-calibration procedures that exploit the information from at least three images let one remove this deformation and produce a 3D reconstruction that is correct up to scale. The limited set of corner or invariant neighbourhood correspondences already yields the necessary data to perform such self-calibration. Hence, self-calibration does not depend on the extraction of the dense correspondences. After such fully automated calibration one knows the camera projection matrices for its different subsequent positions. Once these matrices are available, the 3D model of the observed scene can be produced.

Earlier versions of the system required that intrinsic camera parameters, such as the focal length, remained fixed during image acquisition. But if one has limited *a priori* knowledge about some intrinsic parameters, such as a known pixel aspect ratio or the fact that rows and columns in the images are orthogonal, then others, such as focal length, can be allowed to change.[4] These extensions have been integrated into the system by now. Sagalassos will become an interesting testing ground to validate and test the robustness and quality of such self-calibrating 3D modelling processes.

Figure 8.3 shows a part of the Roman bathhouse that has been modelled in 3D using this passive technique. The top row shows three of six images that were taken

[1] Armstrong et al. (1994).
[2] Tuytelaars et al. (1999).
[3] Koch et al. (1998).
[4] Heyden and Aström (1997); Pollefeys et al. (1998, 1999).

(a)

(b)

Figure 8.4 The passive 3D acquisition approach allows the archaeologists to reconstruct the stratigraphy found at the excavations from the same photographs they already use to document their finds.

of the bathhouse, and then used for the creation of its 3D model. The bottom row shows three views of this model. For the moment only a few such models have been produced, but it is the goal of the project to record in 3D several of the ruins in this way. The same method will be applied to model the Sagalassos landscape. Several images have already been taken along the top of a hill overlooking the excavation site. This has already yielded a global model of the site. In all cases the intrinsic and extrinsic parameters of the camera were unknown.

Under Murale, another type of 3D modelling will be used that is of particular importance to archaeology. Excavations are carried out layer by layer. Although drawings and photographs are made to document the progressing excavation and corresponding stratigraphy, a full 3D representation would be preferable. Hence, one of the goals of the project is exactly this: to build 3D models of the different layers, which can subsequently be virtually scrolled through, with layers being dynamically added or removed, and with visualisation from any viewpoint. Again, the archaeologists have only to take images, as is already current practice. The 3D modelling of the layers is carried out by the University of Leuven, and the visual integration of the 3D layer models is one of the tasks of Brunel University. Figure 8.4 shows two views of the same excavation quadrant, one before and one after a layer had been removed.

Another application that Murale envisages is the use of 3D acquisition technology for virtual or real restoration and *anastylosis*, i.e. the use of 3D shapes of building blocks, sherds, and pieces in general to see how they can fit together. If the building or the artefact to which the pieces belong is of high scientific or artistic value a real restoration can then follow. An example where the passive technique can be applied is the fitting of fractured pillars. Rather than trying to match physically the very heavy pieces with the help of cranes, the reconstructed surfaces can be fitted virtually by the computer. When discussing the active method, a similar application will be the virtual reconstruction of potsherds.

It is interesting to note that the passive method not only produces 3D models of the site, but that it also yields information about where the camera has been and about its settings at these points. If such analysis is performed on a video sequence, virtual objects can be inserted as the projection matrices for subsequent frames are known. The University of Leuven has just started to explore such 'virtual reality' for archaeology. It can be used to give the general public an idea of what the city looked like at different times, by dynamically superimposing 3D CAD reconstructions — of buildings, streets, walls, and so on — onto videos of the site in its current state. Figure 8.5 shows a preliminary example, where a CAD model of the fountain (Nymphaeum) at the upper agora of Sagalassos has been superimposed onto a video sequence showing its current state. The 3D CAD model of the Nymphaeum has been produced and made available by Axell Communications, a Belgian multimedia production company.

Active Method: A 3D Camera

The 'passive' technique has difficulties with untextured parts and the details of complicated shapes. Its geometric precision is limited. These are major problems with objects such as statues or ornaments, which typically consist of untextured stone, but the shape of which should be extracted with high precision. For the 3D shape extraction of such ornamental objects, Eyetronics uses its 'active' ShapeSnatcher system.[5]

'Active' systems bypass the problem of insufficient texture by projecting a pattern onto the scene. The 3D shape is extracted by analysing the displacements/deformations of the pattern when observed from a different direction.[6] Typically, such methods have relied on the projection of single points or lines and on scanning the scene to gradually build a 3D description point by point or line by line. Eyetronics' ShapeSnatcher contains no moving parts as it directly projects and analyses a complete line grid instead. The components needed are a normal slide projector, camera, and computer. The grid, provided on a slide, is projected onto the scene, which is observed from a different viewing angle by the camera. The software then allows extraction of

[5] http://www.eyetronics.com.
[6] See Jarvis (1983) and Besl (1988) for an overview.

Figure 8.5 Frames from the Axell video, with a 3D CAD model of the Nymphaeum superimposed on the actual site. Frames 1 and 2 show the current state of the site, with superimposed model shown in frames 3 to 6.

the 3D shape of the complete patch visible to both. The angle between the viewing and projection directions can be kept pretty low, around 7 to 10°, so that as small a part of the surface as possible is lost to occlusions. With ShapeSnatcher the 3D positions of up to 360,000 points can be extracted from a single image. Thus far, approaches that projected several lines or other patterns simultaneously had heavily depended on the inclusion of a code into the projected structure, which kept the resolution of the extracted shapes low.[7]

Figure 8.6 shows the set-up and a detail of an image from which 3D information can be extracted. The camera is a prototype of a compact and lightweight portable 3D camera.

The result obtained from a single image captured with ShapeSnatcher is not a mere 3D-point cloud, but a connected surface description (the grid in 3D). And this surface is textured as well. In order to extract the surface texture also, the lines of the grid are filtered out. Obviously, an alternative for static objects is to take another image without the grid. This is supported by the software and often done in an archaeological setting, where objects are static and one would like to have the best texture information possible. Although not very conspicuous, the reduction in texture quality obtained by filtering out the lines is an issue.

Figure 8.7 shows the partial 3D reconstruction of a statue of Dionysos that has been excavated at Sagalassos. It has been suggested that it was modelled after Alexander. Such objects obviously cannot be modelled from a single ShapeSnatcher image. Several need to be combined in order to build a complete model. Eyetronics also provides ShapeMatcher software, which supports

(a)

(b)
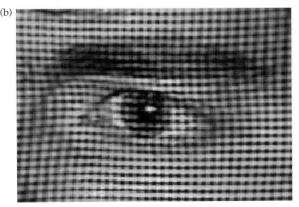

Figure 8.6 (a) Prototype 3D camera. The system consists of a flash projecting a grid and a camera. The camera takes an image from a direction that is slightly different from the direction of projection. (b) A regular square pattern is projected on the scene, as seen in this detailed view. 3D co-ordinates are calculated for all the line intersections, resulting in simultaneous measurement for thousands of points.

---

[7] Boyer and Kak (1987); Vuylsteke and Oosterlinck (1990); Maruyama and Abe (1993).

the knitting of several partial 3D patches. Given the fact that such heavy objects, weighing several tons, need to be modelled, it is important that the acquisition system can be brought to the pieces and not vice versa. It would

(a)

(b)

(c)

Figure 8.7 (a) Statue of Dionysos. (b), (c) Two views of the extracted 3D model.

(a)

(b)

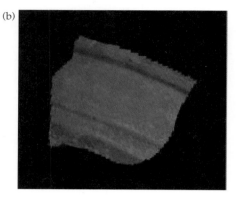

(c)

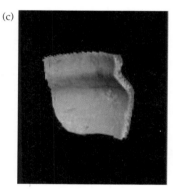

Figure 8.8 (a) A sherd found at Sagalassos. (b) 3D reconstruction, seen from the 'outside', with the external and internal surfaces already aligned. (c) View of the 3D reconstruction seen from the 'inside'.

for instance be difficult to put such a statue into the working volume of a laser scanner, if it would fit in there at all and if the scanner could withstand the pressure.

A special set-up will be produced for potsherds. Such sherds play an important role in archaeology. Although pottery is only rarely recovered in a complete form, ceramic sherds are exceptionally stable over time. They are robust and practically never reused. The outline has to be delineated precisely and the thickness of these parts has to be measured with good precision as well. In this case, speed is of particular importance. Literally millions of sherds are excavated each year in Sagalassos,

of both common ware and fine ware. Although it is obviously not the goal to extract the 3D shapes of all of these, there still is a need to process statistically relevant quantities and to infer their probable silhouette. This is a task of the Technical University of Vienna. It is current practice that experienced archaeologists in the team draw the silhouettes by hand. A disadvantage is that experiments have shown that the drawings of different experts may differ quite substantially. Although some trials with laser-based methods have been carried out in the past, the speed of such automated scanning in the end proved much lower than that of manual sketching, and the trials were aborted for that reason. A one-shot system like ShapeSnatcher may resolve this problem. Figure 8.8 shows preliminary results of acquired 3D shapes for one of the many sherds found on the site. The team at Vienna

(a)  (b)

Figure 8.9 The old bathhouse and surrounding landscape at Sagalassos. (a) View with the original landscape texture. As the zoom is very close, the texture is of insufficient quality. (b) The landscape texture has been replaced by synthetic texture.

are also developing methods to enable the computer to assemble sherds into more complete pots and vessels.

Being able to extract the silhouettes efficiently is important for both common ware and fine ware. In the case of common ware, this typically has regional characteristics. The typology therefore has to be determined on the basis of local finds. Being able to process more pieces yields better statistics and more reference material. Sagalassos became an important centre of fine ware export in Roman times. Here, being able to draw up a good typology has importance also for other excavations, as it can help to date finds in layers were Sagalassos ware was found.

III. Image-Based Texture Synthesis

Only a rather rough model of the landscape (terrain model) will be built. Its resolution will not match that of the 3D models of the ruins, and this is the case for both the geometry and the texture. Visualisation of virtual Sagalassos will require that one can freely navigate through the scene. As one moves from building to building and crosses the terrain in between, a noticeable and disturbing difference in visual quality between the buildings and the terrain will appear. ETH Zurich is developing tools to map realistic texture on the terrain. But this is not the only use of texture synthesis in Murale. The virtually reconstructed buildings (CAAD — computer-aided architectural design — by the University of Graz and Imagination) will need to be covered by textures that emulate the appearance of the original, uneroded building materials. And returning to the landscape, it will need to be covered by texture that simulates the ancient rather than the existing vegetation when the site is visualised as it was in antiquity. This section first describes the basis texture analysis and synthesis approach used for the simpler cases, and then moves on to the use of 'composite textures' to deal with the synthesis of more complicated materials and landscapes.

The Basic Texture Analysis and Synthesis Method

As the first example of the use of synthetic texture, we consider the terrain again. Precise modelling of the landscape geometry and texture all over the tens of square kilometres spanned by the site would cost an enormous

amount of time and memory. Also, such precise modelling is not really required for general visualisation purposes. For this it typically suffices to cover the landscape with a texture that looks detailed and realistic, but that does not necessarily correspond to the real texture on that particular part of the site. There is no need to capture precisely every bush or natural stone. Thus, as a compromise we model the terrain texture on the basis of selected example images of real Sagalassos texture. The terrain model is then covered with similar textures of the right type. Texture synthesis is based on texture models learnt automatically from the example images. A preliminary example of such a procedure is shown in Figure 8.9.

The synthetic vegetation texture (b) that replaces the low-resolution texture of the original terrain model (a) is more in agreement with the resolution of the 3D ruin model that has been integrated into the terrain model (the part of the Roman bathhouse shown in Figure 8.3). The texture model was learnt from an example image of the current vegetation at Sagalassos (thorn-cushion steppe). The texture model is very compact and can be used to generate arbitrarily large patches of this texture.

A multitude of texture descriptions has been proposed in the literature, all with their pros and cons. ETH's basic approach is in line with the co-occurrence tradition, which seems to offer a good compromise between descriptive power and model compactness. Textures are synthesised so as to mimic the pairwise statistics of the example texture. This means that the joint probabilities of different colours at pairs of pixels with a fixed relative position are approximated as closely as possible. Just including all pairwise interactions in the model is not a viable approach and a good selection needs to be made.[8] ETH has opted for an approach that makes a selection so as to keep this set minimal but also brings the complete pairwise statistics of the synthesised textures very close to that of the example textures.[9] Pair type selection follows an iterative approach, where pairwise interactions are added one by one to the texture model, the synthetic texture is each time updated accordingly, and the statistical difference between the example texture and the synthesised texture is analysed to decide which further addition to make. The set of pairwise interactions selected for the model (from which textures are synthesised) is called the *neighbourhood system*. The

[8] Gagalowicz and Ma (1985).
[9] Zalesny and Van Gool (2001).

complete texture model consists of this neighbourhood system and the *statistical parameter set*. The latter contains the joint probabilities for the selected relative pairwise pixel positions, also called *cliques*.

A sketch of the algorithm is as follows.

Step 1 Collect the complete second-order statistics for the example texture, i.e. the statistics of all pairwise interactions. (After this step the example texture is no longer needed.) As a matter of fact, the current implementation does not start from all pairwise interactions, as it focuses on interactions between positions within a maximal distance.

Step 2 Generate an image filled with independent noise and with values uniformly distributed in the range of the example texture. This noise image serves as the initial synthesised texture, to be refined in subsequent steps.

Step 3 Collect the full pairwise statistics for the current synthesised image.

Step 4 For each type of pairwise interaction, compare the statistics of the example texture and the synthesised texture and calculate their 'distance'. For the statistics the intensity difference distribution (normalised histograms) were used and the distance was simply Euclidean. In fact, the colour distributions of the images were added also, where 'singletons' played the role of an additional interaction. The current implementation uses image quantisation with 32 grey levels.

Step 5 Select the interaction type with the maximal distance (cf. step 4). If this distance is less than a threshold, go to step 8 — the end of the algorithm. Otherwise add the interaction type to the current (initially empty) neighbourhood system and all its statistical characteristics to the current (initially empty) texture parameter set.

Step 6 Synthesise a new texture using the updated neighbourhood system and texture parameter set.

Step 7 Go to step 3.

Step 8 End of the algorithm.

After the eight-step analysis algorithm we have the final neighbourhood system of the texture and its statistical parameter set. This model is very small compared to the complete second-order statistics extracted in step 1. Typically only 10 to 40 pairwise interactions are included and the model ranges from a few hundred to a few thousand bytes. Nevertheless, these models have proven effective for the synthesis of realistic-looking textures of a wide variation. Figure 8.10 shows a few examples. The top row textures are images of real building materials (stones) used at Sagalassos. The bottom row shows the results of texture synthesis based on models extracted from the top row images.

In the case of Sagalassos, the method has mainly been used for coloured textures. The examples of Figure 8.10 are a case in point. For the modelling of coloured textures pairwise interactions are added that combine intensities of the different colour bands. The shortest four-neighbourhood system within the colour bands and the inter-band interactions between identically placed pixels were always pre-selected because experiments showed that they are important for the vast majority of the texture classes.

It does not suffice to synthesise separate textures. Where different textures meet clear seams would appear. Hence, a texture-knitting tool was developed. Consider Figure 8.11: (a) shows a composite of rock and grass texture images, both taken at Sagalassos; (b) shows the result obtained with the texture-knitting algorithm. Knitting is based on learning a texture model from the zone around the border between the two textures. Then, new texture is generated in a zone around the border, not necessarily the same, based on the border zone texture model. In the case of a single texture, the seams between separately generated patches can be removed by simply applying that texture's own model near the border.

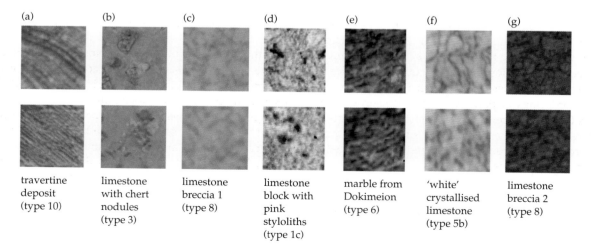

(a) travertine deposit (type 10)

(b) limestone with chert nodules (type 3)

(c) limestone breccia 1 (type 8)

(d) limestone block with pink styloliths (type 1c)

(e) marble from Dokimeion (type 6)

(f) 'white' crystallised limestone (type 5b)

(g) limestone breccia 2 (type 8)

Figure 8.10 Example images of building material (top row) and synthetic textures based on models extracted from these examples (bottom row).

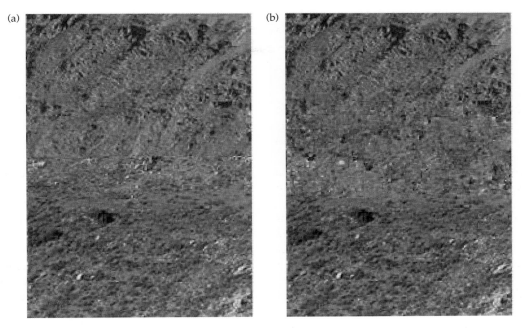

Figure 8.11 Example of texture knitting. (a) Image comprising two types of Sagalassos texture, with rocks on top and grass below. (b) Knitted rock and grass textures.

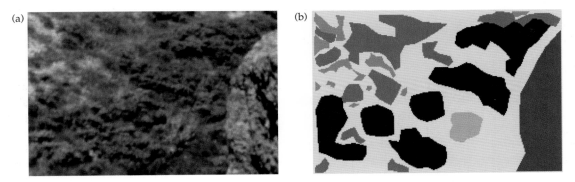

Figure 8.12 (a) An example of modern Sagalassos texture, 'thorn-cushion steppe'. (b) Manual segmentation into basic ground cover types.

Under Murale several extensions to the basic texture synthesis method are planned. First, work is ongoing to extend the texture models to include dependencies of texture appearance on viewpoint and illumination. Another extension aims at dealing with more complicated textures. The basic texture synthesis approach described in this section can handle quite broad classes of textures. Nevertheless, it has problems with capturing complex orderings of patches which themselves show textures (sometimes referred to as micro-textures in the literature). This is why an extension towards a hierarchical approach — the 'composite texture' idea proposed in the next section — is necessary.

Composite Textures

Figure 8.12(a) shows part of the modern 'thorn-cushion steppe' type of landscape found around Sagalassos. It consists of several ground cover types, such as 'rock', 'green bush', 'sand', etc. (Figure 12(b)). If one were to model this composite ground cover directly as a single texture, the basic texture analysis and synthesis algorithm proposed in the last section would not be able to capture all the complexity in such a scene. Therefore, ETH is working on a hierarchical texture-modelling approach, where a scene like this is first decomposed into the different constituent elements, as shown in Figure 8.12(b). The segments that correspond to different types have been given the same intensity (label). This segmentation has been done manually.

Figure 8.13 shows the image patterns corresponding to the different segments. The textures within the different segments are simple enough to be handled by the basic algorithm. Hence, in this case six texture models are created, one for each of the ground cover types (see caption). Also, the map with segment labels (Figure 8.12(b)) can again be considered to be a texture, describing a typical landscape layout in this case. This texture is quite simple again, and can be handled by the basic algorithm. Hence, such layout textures or label textures can be generated automatically as well. The

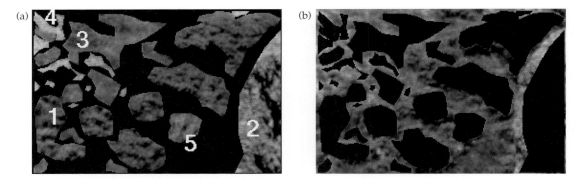

Figure 8.13 (a) Manual segmentation of the Sagalassos terrain texture shown in Figure 8.14: 1 = green bush; 2 = rock; 3 = grass; 4 = sand; 5 = yellow bush. (b) Leftover regions are grouped into an additional class corresponding to transition areas.

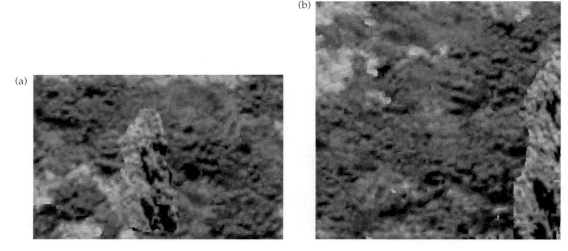

Figure 8.14 Synthetically generated Sagalassos landscape textures: (a) based on a manually drawn label map; (b) based on an automatically generated label map.

idea is then straightforward: generate a *composite texture*, where first a landscape layout texture is generated. Subsequently, the different segments are filled in with the corresponding textures, based on these textures' models. As an alternative, a graphic designer can draw the layout, after which the computer fills out the textures in the segments that he or she has defined, according to their labels. Figure 8.14 shows an example of both procedures.

Note that the right-hand image has been created fully automatically and arbitrary amounts of such texture can be generated, enough to cover the terrain model with never-repeating yet detailed texture. The fact that this approach does not use exact copying of parts in the example images is an important advantage. Once larger areas of texture need to be generated, such copies start to become obvious and therefore distracting for the human observer.

A similar approach can be followed to generate textures of the more complicated types of marbles and limestones that were used as building materials. In a building such as the Nymphaeum, for instance, about ten differently coloured stones were consciously combined to arrive at splendid colour effects. If one wants to recreate

the original appearance of this building and others in the monumental centre of the city, such textures need to be shown with all their complexities and subtleties.

IV. Conclusions and Future Work

We have described the goals of the Murale project. As it has only just started, the purpose of this paper is mainly to indicate in which directions the consortium plans to carry out its further research. In particular, the work of three groups (ETH Zurich, Eyetronics, and the University of Leuven) has been presented in some more detail.

Archaeology poses an interesting mix of challenges to 3D technology. It should be cheap, easy to use and transport, and flexible in the size and type of objects that can be dealt with. Simultaneously the presentation of the results to a wider public will require that the visual quality comes close to or even surpasses what people are becoming accustomed to, i.e. at least game quality. The project hopes to offer such solutions on the basis of photo-realistic modelling tools.

But Murale intends to be about more than generating pretty pictures. It is crucial that its technologies become powerful tools in the hands of the archaeologists themselves. To that end they should in the first place be affordable, so that a large number of excavations can benefit.

Note

The authors gratefully acknowledge support from the EU IST Murale project and also wish to thank Axell Communication for making available its CAD model of the Nymphaeum. Marc Pollefeys gratefully acknowledges support through a postdoctoral FWO grant (Flemish Fund for Scientific Research). Alexey Zalesny is supported as an 'Akademischer Gast' by ETH Zurich.

References

ARMSTRONG, M., ZISSERMAN, A. and BEARDSLEY, P. (1994), 'Euclidean structure from uncalibrated images', in E. R. Hancock (ed.), *BMVC94: Proceedings of the 5th British Machine Vision Conference, 13–16 September 1994, University of York* (Guildford).

BESL, P. (1988), 'Active optical range imaging sensors', *Machine Vision and Applications* 1/2: 127–52.

BOYER, K. and KAK, A. (1987), 'Color-encoded structured light for rapid active ranging', *IEEE Transactions on Pattern Analysis and Machine Intelligence* 9/10: 14–28.

EZZAT, T. and POGGIO, T. (1999), 'Visual speech synthesis by morphing visemes', *AI Memo* 1658 (MIT).

GAGALOWICZ, A. and MA, S. D. (1985), 'Sequential synthesis of natural textures', *CVGIP* 30: 289–315.

HEYDEN, A. and ASTRÖM, K. (1997), 'Euclidean reconstruction from image sequences with varying and unknown focal length and principal point', *1997 Conference on Computer Vision and Pattern Recognition (CVPR '97), June 17–19, 1997, Puerto Rico*: 438–43.

JARVIS, R. A. (1983), 'A perspective on range-finding techniques for computer vision', *IEEE Transactions on Pattern Analysis and Machine Intelligence* 5/2: 122–39.

KOCH, R., POLLEFEYS, M. and VAN GOOL, L. (1998), 'Multi-viewpoint stereo from uncalibrated video sequences', in H. Burkhardt and B. Neumann, *Computer Vision: ECCV '98. 5th European Conference on Computer Vision, Freiburg, Germany, June 1998*, vol. 1 (Berlin and London): 55–71.

MARUYAMA, M. and ABE, S. (1993), 'Range sensing by projecting multiple slits with random cuts', *IEEE Transactions on Pattern Analysis and Machine Intelligence* 15/6: 647–50.

POLLEFEYS, M., KOCH, R. and VAN GOOL, L. (1998), 'Self-calibration and metric reconstruction in spite of varying and unknown internal camera parameters', *Sixth International Conference on Computer Vision … January 4–7, 1998, Bombay, India* (New Delhi and London): 90–5.

POLLEFEYS, M., KOCH, R. and VAN GOOL, L. (1999), 'Self-calibration and metric reconstruction in spite of varying and unknown intrinsic camera parameters', *International Journal of Computer Vision* 32/1: 7–25.

TUYTELAARS, T., VAN GOOL, L., D'HAENE, L. and KOCH, R. (1999), 'Matching affinely invariant regions for visual servoing', *International Conference on Robotics and Automation, Detroit, May 1–6*: 1601–6.

VUYLSTEKE, P. and OOSTERLINCK, A. (1990), 'Range image acquisition with a single binary-encoded light pattern', *IEEE Transactions on Pattern Analysis and Machine Intelligence* 12/2: 148–64.

ZALESNY, A. and VAN GOOL, L. (2001), 'A compact model for viewpoint dependent texture synthesis', *3D Structure from Images: SMILE 2000: Second European Workshop on 3D Structure from Multiple Images of Large-Scale Environments, Dublin, Ireland, July 2000*, Lecture Notes in Computer Science 2018 (New York and London).

CHAPTER 9

Three-Dimensional Laser Imaging in an Archaeological Context

Andrew M. Wallace

I. Introduction

In recent years the considerable advances in processing power, memory density, and special purpose hardware have led to the development of increasingly realistic graphical images rendered from 3D model environments. Developers of computer games can now simulate real (car, flight simulators) and imaginary (dungeons, alien planets) scenarios which are ever more pleasing to the eye, and film and TV technicians can now create virtual and mixed media effects that appear ever more realistic without detailed examination. In the majority of cases, these simulations are based on 3D polygonal models that are then rendered with different material and illumination parameters and from different viewpoints to produce an almost infinite number of possible visual images. The 3D models may simply be the product of a designer's imagination, but increasing use is being made of 3D imaging technology to recreate near-exact replicas of original objects and scenes that can be encoded in computational format. This has important uses in industry, for example in quality control or examining the aerodynamic properties of an aeroplane wing or a car body, but has also assumed increasing importance in the reconstruction and recreation of archaeological sites and artefacts, as discussed in this volume and elsewhere.[1]

In this chapter we consider the strengths and weaknesses of techniques of 3D laser imaging based on triangulation and time of flight. Laser scanning by triangulation is a relatively mature technology, and it is now possible to build or purchase well-characterised, accurate, and fast systems, which can create faithful 3D datasets from a reasonable range of surface materials. However, they suffer from a number of drawbacks that restrict their universal applicability. The geometry of the sensor means that there is an inevitable trade-off between the scale of the object measured, the stand-off distance, and the accuracy of measurement. For surfaces with concavities or other intricate surface detail, there is the possibility of occlusion due to the necessary separation of viewpoint between the viewing camera and the laser projector. Further, the data may be corrupted by poor and false returns caused by variable material reflectance. Specular and transparent materials cause particular difficulties.

In a time-of-flight system, distance is measured by measuring the time for a focused laser beam to impact and return from the surface of interest. Knowing the speed of light, the distance is computed and a 3D image can be created if the laser beam is scanned across the target. Although this is an attractive option, eliminating the occlusion problem in particular, there are still outstanding problems with the current methodologies based on modulation of a laser signal. The time resolution required to measure range to sub-millimetre accuracy is very difficult to achieve, and in many cases the measured time is also affected by the magnitude of the returned signal, and hence the reflectance, surface angle, and range of the target. In addition to reviewing the current technologies, we present in this chapter the development of a new approach to time-of-flight laser imaging based on time-correlated single photon counting. This has a number of advantages over previous methods, in particular the ability to resolve very short time, and hence distance, differences, and sensitivity to low light returns, making it possible to make 3D measurements on distant, poorly reflecting, or even primarily transparent surfaces.

To illustrate these techniques, we have been fortunate to have access to original or copies of artefacts from the National Museum of Scotland, Edinburgh, and the Hunterian Museum in Glasgow. Therefore we present examples of 3D images and models acquired by both triangulation and time of flight. We also illustrate how real and synthetic data can be combined easily to create an augmented, rather than a simply virtual, scene.

II. 3D Archaeological Imaging by Passive and Active Methods

In an archaeological context, considerable use has been made of passive techniques, i.e. the reconstruction of 3D models from two or more images of the scene or object under consideration. In fact, in some cases 3D models have been built by the study of drawings, photographs, and plans using manual measurement as input to a computer-modelling package which could then create an impression of the original environment.[2] Photogrammetric techniques offer a more automated means

---

[1] Forte and Siliotti (1997); Ogleby (1999); Van Gool et al. (Chapter 8 in the present volume).

[2] Bloomfield and Schofield (1996).

to create 3D data from two or more intensity images acquired from different viewpoints. Provided corresponding points or features can be identified within these images then it is possible to use the known imaging parameters of the cameras to recover 3D information from the scene. However, it is not easy for a computer system to identify corresponding points. Sometimes the correct correspondence between two images is chosen but a difference in viewpoint or illumination, or even simple sensor noise, can make the observed feature appear different. This can lead to errors in determination of the exact positions of the feature in the 2D images, and hence to inaccuracies in the 3D geometry. In other cases, the wrong correspondences are made, leading to gross errors. To obviate these problems it is common to place calibrated targets on the surfaces under examination that can be clearly identified in each image. For example, this was the process used by Ogleby[3] in his reconstruction of Ayutthaya. Although it is still a time-consuming process, there are standard software packages that can be used to calibrate the cameras and recreate the 3D positions from the corresponding targets. Recently, researchers in computer vision have looked at the recovery of 3D data using uncalibrated systems.[4] Apart from the theoretical interest, this has important uses for 3D navigation by mobile robots (whose position is not known) or 3D reconstruction from archival photographs (where little may be known), to name but two examples. These techniques may well be useful in an archaeological context but would not achieve the same accuracy as true photogrammetry and rely heavily on known or assumed constraints, for example parallel and orthogonal structures in buildings.

In an active 3D sensing system, an energy source, usually a laser, is projected onto the scene, which is viewed by one or more cameras at a displaced viewpoint.[5] In a triangulation system, a point source (spot) may be scanned in two dimensions, a stripe may be scanned in one dimension, or a complete 2D pattern/texture may be projected onto the scene, eliminating the necessity for any scanning. In general, there is a compromise between speed of scanning, ease of use, and the required accuracy, and the scanned stripe is the technique which has found most acceptance, in applications ranging from car body parts to museum figurines. For example, the system developed by the Canadian Research Institute has been used to acquire both colour and 3D data to sub-millimetre resolution simultaneously from a wide range of museum and cultural objects.[6] However, it is difficult to build a portable triangulation system that can be used to create 3D images of large structures. For that reason, Robson Brown et al.[7] chose to use a time-of-flight system to obtain a reasonably accurate model of an Upper Palaeolithic rock

shelter in south-west France. This was imported into a proprietary 3D modelling package and viewed under different illumination conditions to highlight detail.

The majority of the applications of 3D imaging technology to archaeological artefacts have been to create high-quality digital records, or to be used as part of a virtual museum or gallery, which can be viewed readily by a remote student. As in industry, captured 3D models can be used to create replicas using numerically controlled machines, and if an original artefact is broken it is possible to use computer procedures to try to reassemble the pieces either in the virtual environment or by automated means.[8] There appear to have been relatively few examples of the use of 3D imaging or active projection to classify or otherwise augment our knowledge. One such example is the use of shadow stereo imagery to enhance images of incised documents.[9] Another example has been the analysis of the reflectance distribution function from vases impacted by a He–Ne laser to identify the nature of the gloss on the surface.[10] In passing, we note that polarisation analysis can provide an even richer analysis of the material type and reflectance properties.[11]

III. 3D Imaging by Triangulation

Triangulation is a mature and reliable technique for the acquisition of 3D image data from opaque surfaces. The basic geometry of the triangulation technique shown in Figure 9.1 is implemented in the flatbed scanner illustrated in Figure 9.2, which was constructed at Heriot-Watt University and has been used extensively over the last few years. In Figure 9.1, A and B are nodal points on the projection axis of a laser beam (AC) and the optical axis of a camera (BC) respectively. In Figure 9.2, the laser beam is not visible but is projected through the optics visible on top of the platform to form a plane of light projected vertically downwards onto the

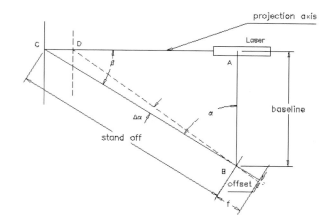

Figure 9.1 The basic geometry of a triangulation scanner.

[3] Ogleby (1999).
[4] Cipolla and Robertson (1999).
[5] Clarke (1998).
[6] Baribeau et al. (1996).
[7] Robson Brown et al. (2001).

[8] Ritchie et al. (1999).
[9] Brady et al. (Chapter 2 in the present volume).
[10] Maniatis et al. (1993).
[11] Wolff (1990); Liang et al. (1999).

Figure 9.2 A photograph of a conventional flatbed scanner used at Heriot-Watt University.

metallic block held by the researcher. The sensor has three cameras: the one in the middle is actually used for capture of coincident intensity data but either displaced camera can be considered as equivalent to the imaging system in Figure 9.1. The baseline separation (BA) is the shortest distance between the optical centre of the camera and the projection axis of the laser beam, and the stand-off distance is defined as the distance from the intersection of the projection and optical axes (C) to the optical centre (B). A change in the distance of the object from the laser nodal point at D results in a ray of light entering the camera at an angle $\Delta\alpha$ to the camera axis measured by the observed offset position of the light ray on the sensor plane. In general, the geometry of a conventional triangulation system is fixed; the angle β and the baseline separation cannot be dynamically changed. As shown, the triangulation system acquires a depth profile of the block: to acquire a full-depth image either the laser beam must be scanned or the block moved. In the system shown a linear translation stage moves the block through the beam and data capture is synchronised to the lateral motion, and a 3D image is acquired automatically.

There are several limitations and problems with such a system. There is an inevitable trade-off between the scale of the object measured, the stand-off distance, and the accuracy of measurement. In effect, this means that it is necessary to use different sensors for different sizes of object, and as the sensor plane is usually sampled at discrete points the absolute accuracy of measurement is also scaled. Second, there is the occlusion problem: for a depth point to be acquired it must be visible to both the laser projector and the displaced camera, and hence any concavities in an object may not be scanned. This is why there are two displaced cameras in Figure 9.2, since it increases the chance of surface visibility by at least one camera and the laser. Of course reduction of the baseline in Figure 9.1 reduces occlusion but also reduces

accuracy. One solution is to use a continuously variable geometry so that this trade-off can be adapted to suit the object being scanned.[12] A further problem can occur with poor and false returns due to variable material reflectance and multiple reflections; the latter is most usually evident when scanning metallic surfaces that have significant concavities. Polarisation optics can also be used to reduce this problem, since the characteristics of the first reflection are predictably different from subsequent reflection and can be filtered out.[13] However, a simpler solution (if feasible) is to spray the object with fine white powder! Finally, it should be emphasised that a system of the type shown in Figure 9.2, though fully automated, can acquire a depth image from only one point of view. If a complete, all-round 3D model of an object is necessary then some further rotation of either object or scanner is required.

Figure 9.3 shows an example of a commercial triangulation system which adopts an engineered approach to acquire full 3D data. This uses a co-ordinate measuring arm with six degrees of freedom; each rotating joint has an encoder which measures the precise angle of the joint, and hence the position of the probe tip in 3D Cartesian space can be computed. The black triangulation scanning head is mounted at the end of the arm in close proximity to the touch probe and has an enclosed laser and sensor. To scan all round an object the arm is manipulated, and at each position a section of the surface is scanned. The final 3D dataset is just the conjunction of all these separate scans and is effectively a 'cloud of points' in (x, y, z) format. Nominally, this maintains an accuracy of approximately 0.1 mm over a working volume of $0.5 \times 0.5 \times 0.5$ m, although in practice the final accuracy is dependent to a large extent on how the arm is moved by the operator. If the path between different scans is complex then there is more chance that the relative positions will be inaccurate when the separate segments are merged, because of inaccuracies in the robot kinematics.

Figure 9.4 shows a detail of the result of a triangulation scan of the head that is also shown in Figure 9.3. This bust is a copy of a Roman artefact found in Bearsden, Dunbartonshire, generally identified as the goddess Fortuna from its location in the bathhouse, and currently held by the Hunterian Museum, Glasgow. The original data are a cloud of points, but the picture depicts a few of the more than 240,000 polygons which have been meshed from the original point cloud data. The polygonal head is viewed in profile, so that it is possible to see on the right, from bottom to top, the chin, the mouth, and part of the broken nose and cheek bone. In Figure 9.5 the head has been imported into a virtual environment to illustrate how accurate 3D representations of real objects can be viewed in an appropriate setting, in this case a supposed study. The head has been shaded and textured with a stone surface, although this can be chosen at will, or a simultaneous photograph can be taken and mapped

[12] Clark et al. (1998).

[13] Wallace et al. (1999).

Figure 9.3 A photograph of a commercial triangulation scanner mounted on a measurement arm, with a copy of a Roman bust of the goddess Fortuna.

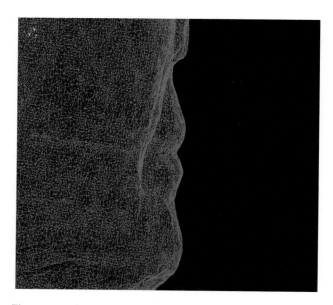

Figure 9.4 A detail from the nose, mouth, and chin of the polygonal mesh formed from the scanned head of Fortuna.

onto the 3D surface. All the other objects in the room are modelled, with the particular exception of the mug, which is also scanned. However, the subsequent processing of the mug was different: a set of higher order surfaces was fitted to the scanned data to give a very compact data structure, although this does have the effect of creating artificially smooth surfaces. Paradoxically,

Figure 9.5 A virtual scene. The desk and other room furniture are all simple models having only a few polygons. The Roman head has been scanned and meshed to give ~240,000 polygons. The cup has been scanned, then fitted by a number of higher-order surfaces, then re-polygonised for display.

the actual viewed image has reconverted the compact representation to polygons for display.

IV. 3D Imaging by Time of Flight

A time-of-flight (TOF) 3D depth measurement system measures the time taken for light (usually from a laser) to travel to and from a target along the line of sight of a common transmitter–receiver optical axis. The beam is along a single fixed axis, so a 2D scanning motion is required to build a 3D image. There are two immediate advantages of the TOF technique. First, there is no occlusion because there is no separation between the source and receiver. As an adjunct to this it is unlikely that a secondary reflection will traverse the same path back to the receiver via the first surface, and this may in any case be outwith the time (distance) window of the sensor. Second, the scale problem is greatly reduced, or alternatively the dynamic range of the system greatly increased, because it is possible to 'point and shoot' the camera at targets over a wide range of distances. This assumes, of course, that the detector can correctly determine the signal return above noise and interference. Unlike interferometric measurement systems that use specially reflecting targets, such as corner cubes, or multiple view techniques, such as photogrammetry, which often require special surface targets, the goal is to measure directly from an uncooperative surface without manual intervention. As the speed of light is $\sim 3 \times 10^8$ m per second it is a major challenge to resolve the short time difference required. For example, to gain 100 μm accuracy of path length, one must resolve time differences of the order of 0.33 ps. To measure this elapsed time, the laser beam is modulated; there are several examples of research prototypes and commercially available TOF systems that use amplitude, frequency, or pulsed modulation.[14]

---

[14] Gagnon (1997); Palojarvi et al. (1997); Kacyra et al. (1999).

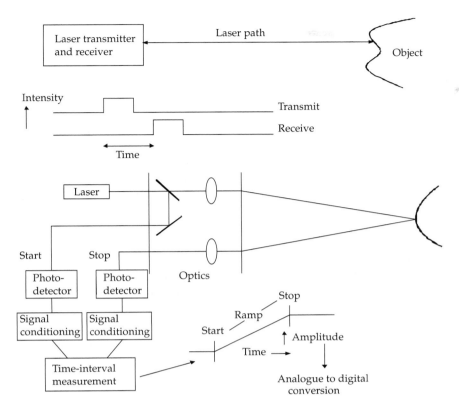

Figure 9.6 A schematic diagram of a conventional pulsed time-of-flight ranging system.

The principal components of a pulsed TOF system are illustrated schematically in Figure 9.6. The collimated laser beam is directed towards the surface of interest, and the reflected signal acquired by a receiving optical system. Since the beam traverses the path length twice, the distance to the surface can be measured as $(1/2)c\tau$, where c is the speed of light and τ the elapsed time, which in this instance is the separation of the positive edges of the transmitted and received pulse. In the lower part of the figure, the laser signal is split between a start and a stop photodetector. The start signal is conditioned and used to trigger a ramp or alternatively start a digital counter that is the datum for the elapsed time measurement. The received signal from the target is detected by the same or another photodiode and conditioned to stop the ramp or counter. Hence, the signal amplitude or the total number of pulse counts provides a measurement of the time elapsed and hence the distance to the target. As highlighted above, a scanning system is required to build a complete 3D image. If the object is small an XY-translation stage can be used similar to the single axis stage shown in Figure 9.2; otherwise the laser beam can be scanned by a mirror system or the whole sensor head moved on a pan-and-tilt head. An amplitude modulated TOF system is conceptually similar. The amplitude of the beam is usually modulated sinusoidally at a fixed frequency so that a measurement of phase shift between the transmitted and received signal provides the equivalent measure of elapsed time. The majority of FM systems have used the self-mixing effect, in which an interference between an internal (to the sensor) cavity and an external (between sensor and object) cavity causes modulation of the output power that is a function of the modulation frequency and distance to target. If the frequency is modulated in a controlled manner, then the range can be computed from the observed output power fluctuations.

Existing TOF systems can measure range to accuracies of the order of 0.1–1.0 mm, but this is usually dependent on favourable operating conditions. First, there must be a strong return signal. In practice, that is a function of the surface reflectance and distance to the target. This is further complicated by the dependence of the measured range on the intensity of the returned signal in all of the modulated techniques. For example, consider the simplest case of a pulsed system in which the time is recorded when the leading edge of the pulse crosses a fixed threshold. If the return signal is reduced in amplitude then it will cross that threshold later even if the pulse shape is unchanged. Hence the range measurement is systematically altered, so that highly reflecting surfaces appear closer than those of low reflectance. Although modulated TOF systems will inevitably include signal processing software to reduce this kind of effect, it proves difficult to eliminate it completely. The shape (rise time, full width half maximum) of the laser pulse is a critical factor, as is the response time of the photodetector. In the example of Figure 9.6, the linearity, or at least calibrated repeatability of the ramp, and the resolution of the ADC, which converts the ramp amplitude to a digital measurement, must all be optimised. Finally, TOF systems are much more sensitive to temperature variations than triangulation systems, although this can be alleviated either by temperature control or by use of a reference channel for calibration.

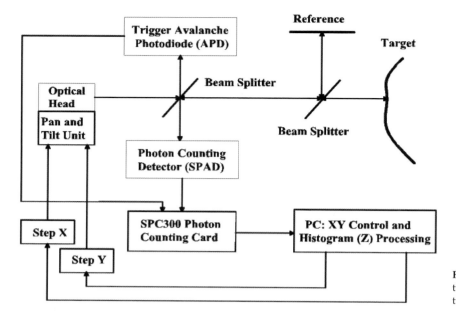

Figure 9.7 A schematic diagram of the time-of-flight ranging system based on time-correlated single photon counting.

V. Time of Flight Using Photon Counting

The perceived limitations of the existing approaches to 3D imaging based on either triangulation or TOF reviewed in Sections III and IV have led us to develop a new approach to TOF ranging based on time-correlated single photon counting (TCSPC). In brief, the desire was to develop a depth measurement technique that could operate at short (<1 m) to long range (>100 m), from both co-operative and non-cooperative surface materials, and at sub-100 μm accuracy. In this section we shall describe the principal features of our new approach and illustrate an example of a depth image acquired from another Roman artefact. Although the system was built originally to operate at an optimum stand-off distance of 25 m and was more typically used to measure aircraft or automotive parts, it was used successfully for the more ancient artefacts under consideration here.

Using the TCSPC technique, the laser is pulsed at a high repetition frequency, but, rather than detect and compare the time separation of a target and reference pulse, the detector has single photon detection sensitivity. There are two significant advantages: first, the detector is capable of sub-picosecond timing resolution, and, second, it does not require a high laser return amplitude, as it responds to a single quantum of light. When compared with the previous modulated methods for 3D measurement based on time of flight, the approach offers very accurate time (and hence distance) resolution, very high sensitivity, and the ability to cope with variation in amplitude of the reflected return of several orders of magnitude.

Figure 9.7 shows a schematic diagram of the TOF-TCSPC distance measurement system which is described fully by Massa et al.[15] The optical head shown in Figure 9.8 is mounted on a pan-and-tilt unit driven by X and Y stepper motors to give an angular accuracy

[15] Massa et al. (1998, 2002).

Figure 9.8 A photograph of the TOF-TCSPC optical sensing head.

of ~5 arc seconds, corresponding, for example, to a transverse resolution of ~25 μm at a distance of 1 m. Optical shaft encoders give the precise angular orientation of the two axes. An IBM-compatible PC is used to synchronise the stepper motors, the driver for the laser mounted in the optical head, and the photon counting detector. The laser source is a passively Q-switched AlGaAs laser diode (developed at the Ioffe Institute, St Petersburg) which emits 10–20 ps pulses of energy of ~10 pJ at 850 nm, with a repetition frequency of up to 25 MHz. A fraction of the source laser signal (~2 per cent) is split off and directed towards a conventional analogue avalanche photodiode (APD), which provides a trigger signal to start the timing system. The remaining laser signal is directed towards the target. The scattered return signal is collected and monitored using a photon counting detector, a single photon avalanche diode (SPAD) which provides a signal to the SPC300 timing system. The time interval is then recorded before the system is reset. The SPAD is an important element of our system: the response consists of a fast rise and

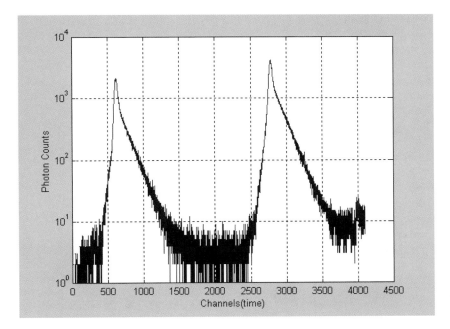

Figure 9.9 An example of raw histogram data collected from the TOF-TCSPC sensor.

a slow wavelength-independent exponential tail. We have used devices developed at the Politecnico di Milano which have a particularly high specification, although commercial alternatives are now available from a number of suppliers. To faithfully record the target return pulse, the detection probability must be sufficiently low (typically <5 per cent) to minimise any multiple photon events which will distort the statistical distribution of photon arrival times. By repetitively pulsing the laser, many photon timing events are observed and a histogram can be constructed which accurately represents the probability of photon arrival times and corresponds to the optical pulse shape in the time domain. In practice, to minimise the effect of drift in the timing electronics, a reference reflecting channel is added, a fixed length optical fibre, which causes a second pulse to appear on the recorded histogram, as shown in the typical example of Figure 9.9. Accurate distance measurements are made by processing the histogram data to measure the separation of the peaks.[16] Background light can be filtered out by use of a wavelength selective filter in front of the photon counting detector, and this is further aided by the finite timing window for the distance measurement and the narrow field of view of the receiving optics.

The raw data collected from the TOF-TCSPC system is a histogram showing the number of detected photons (vertical axis) versus time (horizontal axis). The horizontal axis is defined as sampled channels; in this example each channel corresponds to 2.44 ps. Each time a photon is returned from either target or reference, the count rate in the bin that has the measured time value is incremented by one. The two peaks correspond to the accumulated target (left) and reference (right) single photon returns. The number of photons collected depends on the laser repetition rate, the acquisition time, and the properties of the optical system and target surface. The average photon count rate is maintained typically at 5 per cent or less so that the chance of more than one return within the collection window is small. We also apply a statistical correction factor to allow for the still finite probability of more than one photon, which would, if uncorrected, lead to a bias towards early events. The number of counts in each channel c_i follows a Poisson probability distribution with mean and variance μ_i, where μ_i is the number of observed counts in channel c_i. The histogram also contains a few dark and stray photon counts, but because of the wavelength, temporal, and spatial filtering this is relatively small.

We have investigated a number of different approaches to the processing of the histogram data to get a more accurate estimate of the time lag between the reflected target and reference signals in the histogram[17] either using the raw data or the auto-correlation function. Considering the histogram as a probabilistic density function whose exact functional form is unknown, a nearest representation or 'operating model' has been used to describe the underlying true distribution. This has the form:

$$f(c_i) = \left[h(c_i, \gamma) \left\{ 1 + \sum_{k=1}^{m} a_k \cdot p_k(c_i) \right\} \right] \Big/ \beta$$

where $h(c_i, \gamma)$ is a symmetrical **Loren**tzian function of width γ and $p_k(c_i)$ are polynomial functions, weighted by constants a_k, selected from the first m Hermite or Laguerre polynomials. The product of these gives a model that is asymmetric and general enough to fit the observed form of the received data.

---

[16] Umasuthan et al. (1998).

[17] Ibid.

The performance of the TCSPC-TOF system has been assessed in order to determine the measured longitudinal (z) and transverse (x, y) spatial resolution, single point measurement time, and long-term stability. The longitudinal spatial resolution is determined by the accuracy of the timing and subsequent data processing. Using a micropositioner to verify the ground truth on targets at distance of the order of 1 m, we have achieved absolute measurement of range to an accuracy and repeatability of 15 μm. Transverse resolution is determined by the spot size; for the current set-up the 1/e spot size is 60 × 170 μm at a distance of 2 m and ~400 μm at a distance of 13 m. As currently employed on a pan-and-tilt head, the transverse accuracy is also limited by the precision of the shaft encoders and mechanical drive; this is specified accurate to 5 arc seconds, comparable to the spot size, for example 48 μm at 2 m. It has proved difficult to measure ground truth in this latter case, but we would estimate the measurement in three dimensions to be better than 0.1 mm.

Of course a key element of our approach is the ability to measure directly from the surface of objects rather than rely on reflective targets. Our evaluation of the technique has included measurement from objects of less than 5 cm to about 25 m in size. We have recently extended the range of our evaluation to include measurement at even longer distances and to include the imaging of transparent objects.[18] As the technique relies on the detection of very small quanta of light we are able to use an eye-safe class 1 laser, currently at a wavelength of 850 nm which is not visible to the human eye. A conventional CCD camera is also included in the optical head of Figure 9.2 to enable focusing of the sensor on the target. There is a balance to be struck between the accuracy and time of measurement, for example an accuracy of measurement of 15 μm was made with ~10^6 integrated photons, which corresponds to ~1 s measurement time, worsening to 150 μm at 10^4 integrated photons, which could be accumulated in 0.01 s. We are currently working on new approaches to process the histogram data at low photon counts which will improve this balance, allowing faster measurement at equivalent accuracy, or imaging of more difficult or distant surfaces at slower rates.

Returning to the archaeological context, we have used the TOF-TCSPC sensor primarily to obtain depth images of medium-scale objects of the order of a metre across. In fact, the example in Figure 9.10(a) is less than a metre, about 40 cm across. This is a cast which was borrowed from the National Museum of Scotland in Edinburgh; it dates from about (AD) 1900, but the original Roman frieze was found near Jedburgh in the second century AD. The depth data are shown in the form of a pseudocolour image (b), in which the 'hotter' colours are nearer to the sensor. We have not established ground truth by other means, and the picture provides only an illustration, but this is a dense depth image in the form of a point cloud that appears to reproduce the relief of the frieze with good accuracy.

[18] Wallace et al. (2000).

(a)

(b)

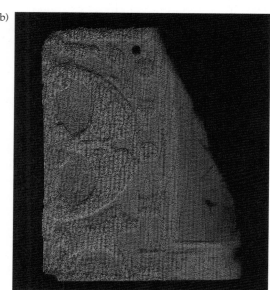

Figure 9.10 (a) A picture and (b) a depth image of a cast of a Roman frieze, scanned using the TOF-TCSPC technique. A pseudocolour spectrum is used in which red indicates points nearest to the camera and blue indicates points furthest from the camera.

VI. Conclusions

In this paper we have given a brief description and discussed the relative advantages and disadvantages of the two principal approaches to acquire dense 3D depth data directly from object surfaces, and reviewed their use in an archaeological context. Currently, scanning by triangulation is by the far the most popular technique and can be used to acquire single viewpoint or complete 3D models from objects of relatively small scale, typically up to 2–3 m in volume, achieving best accuracies of the order of 1 mm or less. If the objects have significant surface complexity, or variations in surface reflectance, there may be difficulties, but in general there are engineered approaches to lessen these problems.

Until now, time-of-flight systems have tended to be used for scanning larger objects, such as building façades, for two major reasons. First, it is difficult to get equivalent accuracy to triangulation systems at smaller scale with the pulsed, amplitude, and frequency modulated techniques because of the difficulty of very short time measurement. Second, it is very difficult to build a triangulation system to get accurate distance resolution at larger scales because of the dependence of the depth resolution on the baseline separation between transmitter and receiver, the sensor sampling interval, and other geometrical factors. However, we have developed a new approach to time-of-flight imaging using photon counting that has proved useful in imaging both small and large objects, achieving excellent depth resolution from a wide variety of objects and surfaces.

We have illustrated the paper with some examples from local museums. A study of the literature shows an increasing use by historians of 3D visual techniques, primarily for visualisation and display, locally and remotely, the 'virtual museum'. Further scope exists to use the reverse engineering capability of 3D scanning, i.e. to create a CAD replica of a real museum artefact and use it to create replicas in a new or similar material. The use of 3D metrology to better understand the artefacts, for example to analyse a surface to discover how it has been worked or to discriminate between different interpretations of faint lettering, seems to be relatively unexplored, and would provide fertile ground for new research.

Note

The author would like to acknowledge the contributions made to this work by J. Massa, M. Umasuthan, J. Clark, S. Pellegrini, G. Smith, P. Csakany, G. Buller, and A. Walker. We are also grateful to A. Sheridan at the National Museum of Scotland and L. Keppie at the Hunterian Museum, Glasgow, for allowing us to scan artefacts of interest.

References

BARIBEAU, R., GODIN, G., CORNOYER, L. and RIOUX, M. (1996), 'Colour three dimensional modelling of museum objects', *Imaging the Past*, British Museum Occasional Paper 114 (London): 199–206.

BLOOMFIELD, M. and SCHOFIELD, L. (1996), 'Reconstructing the treasury of Atreus at Mycenae', *Imaging the Past*, British Museum Occasional Paper 114 (London): 235–43.

BRADY, M., PAN, X.-B., SCHENK, V., TERRAS, M. and MOLTON, N. (2003), 'Shadow stereo, image filtering, and constraint propagation' [Chapter 2 of the present volume].

CIPOLLA, R. and ROBERTSON, D. P. (1999), '3D models of architectural scenes from uncalibrated images and vanishing points', *10th International Conference on Image Analysis and Processing, Venice*: 824–9.

CLARK, J., WALLACE, A. M. and PRONZATO, G. (1998), 'Measuring range using a triangulation sensor with variable geometry', *IEEE Transactions on Robotics and Automation* 14/1: 60–8.

CLARKE, T. (1998), 'Simple scanners reveal shape, size and texture', *Optics and Laser Europe* 4: 29–32.

FORTE, M. and SILIOTTI, A. (1997), *Virtual Archaeology: Recreating Ancient Worlds* (New York).

GAGNON, E. (1997), 'Laser range imaging using the self-mixing effect in a laser diode', *IEEE Transactions on Instrumentation and Measurement* 48/3: 693–9.

KACYRA, B., DIMSDALE, J. and BRUNKHART, M. (1999), *Integrated System for Quickly and Accurately Imaging and Modeling Three Dimensional Objects*, Cyra Technologies, Inc., US Patent 5,988,862.

LIANG, B., WALLACE, A. M. and TRUCCO, E. (1999), 'Visual classification of materials using the Stokes vector', *Polarization and Color Techniques in Inspection: Proceedings of SPIE, the International Society for Optical Engineering*: 224–333.

MANIATIS, Y., ALOUPI, E. and STALIOS, A. D. (1993), 'New evidence for the nature of the Attic black gloss', *Archaeometry* 35/1: 23–34.

MASSA, J. S., BULLER, G. S., WALKER, A. C., COVA, S., UMASUTHAN, M. and WALLACE, A. M. (1998), 'A time-of-flight optical ranging system using time-correlated single photon counting', *Applied Optics* 37/31: 7298–304.

MASSA, J. S., BULLER, G. S., WALKER, A. C., SMITH, G., COVA, S., UMASUTHAN, M. and WALLACE, A. M. (2002), 'Optical design and evaluation of a three-dimensional imaging and ranging system based on time-correlated single-photon counting', *Applied Optics-LP* 41/6: 1063–70.

OGLEBY, C. L. (1999), 'From rubble to virtual reality: photogrammetry and the virtual world of ancient Ayutthaya, Thailand', *Photogrammetric Record* 16/94: 651–70.

PALOJARVI, P., MAATTA, K. and KOSTAMOVAARA, J. (1997), 'Integrated time of flight laser radar', *IEEE Transactions on Instrumentation and Measurement* 46/4: 996–9.

RITCHIE, J. M., DEWAR, R. D. and SIMMONS, J. E. L. (1999), 'The generation and practical use of plans for manual assembly using immersive virtual reality', *Journal of Engineering Manufacture* 213: 461–74.

ROBSON BROWN, K. A., CHALMERS, A., SAIGOL, T., GREEN, C. and D'ERRICO, F. (2001), 'An automated laser scan survey of the Upper Palaeolithic rock shelter of Cap Blanc', *Journal of Archaeological Science* 28: 283–9.

UMASUTHAN, M., WALLACE, A. M., MASSA, J. S., BULLER, G. S. and WALKER, A. C. (1998), 'Processing time-correlated single photon data to acquire range images', *IEE Proceedings: Vision, Image and Signal Processing* 145/4: 237–43.

VAN GOOL, L., POLLEFEYS, M., PROESMANS, M. and ZALESNY, A. (2003), 'Modelling Sagalassos: creation of a 3D archaeological virtual site' [Chapter 8 in the present volume].

WALLACE, A. M., CSAKANY, P., BULLER, G. S. and WALKER, A. C. (2000), '3D imaging of transparent objects', in M. Mirmehdi and B. Thomas (eds.), *BMVC2000: Proceedings of the 11th British Machine Vision Conference, 11–14 September 2000, University of Bristol*: 466–75.

WALLACE, A. M., LIANG, B., CLARK, J. and TRUCCO, E. (1999), 'Improving depth image acquisition using polarized light', *International Journal of Computer Vision* 32/2: 87–109.

WOLFF, L. B. (1990), 'Polarisation based material classification from specular reflection', *IEEE Transactions on Pattern Analysis and Machine Intelligence* 12/11: 1059–71.

CHAPTER 10

Movements of the Mental Eye in Pictorial Space

Jan J. Koenderink

Introduction

When the human observer is confronted with a straight photograph of some object, the observer can either look *at* the photograph or the observer can look *into* the photograph. The difference is crucial, for in the former case the observer is aware of the photograph as a physical object in physical space, whereas in the latter case the observer is aware of a pictorial object in pictorial space and — simultaneously in subsidiary awareness — of the photograph as a physical object in physical space. To be able to be aware of the photograph as a physical object is a competence humans share with the higher animals, but the ability to be aware of a pictorial space is probably singularly human. It is debatable whether even the non-human primates are capable of it. Pictorial space is categorically different from physical space in that it doesn't exist outside of an observer's awareness. It is a thread of consciousness, a purely mental object. The photograph as a physical object is a roughly planar sheet, covered with pigments in a certain simultaneous order. When suitably illuminated and viewed this distribution of pigments can become the momentary object of one's vision. Such visual objects are categorically different from pictorial objects in pictorial space, though.

For the genesis of pictorial objects the particular ontology of the photograph is immaterial: that is to say, it is not relevant that the photograph is the result of a process in which at some earlier time some person (the photographer) exposed a photosensitive sheet which happened to be irradiated with the optical image of a physical scene, and so forth. The earlier existence of a physical scene that figured causally in the present existence of the photograph in no way 'explains' the pictorial object and scene. This is a crucial observation that seems rarely well understood. Most people find it natural to assume that one understands what is in a certain holiday snap shot *because* the leaning tower of Pisa actually existed (at least at the time of exposure of the photograph). However, there exists really no ground for such a cheerful assumption. On closer analysis it might turn out to be the case that what was taken for a photograph was really the product of a particularly dextrous artist in the super-realist style, or that it was mere fungus growth (in a peculiar pattern) on a piece of dried wood pulp after all; one could multiply possible explanations for the existence of the purported 'photograph' endlessly. The point is that only the simultaneous order of pigments co-determines the pictorial

scene, not the 'history' connected with the photograph. Unicorns don't populate physical space, but are nothing remarkable in pictorial space. I say 'co-determine', because there is obviously 'the beholder's share'[1] since pictorial spaces exist only in the minds of beholders. It is well known that pictorial scenes are often mainly determined by the order of pigments (as when several people readily agree on what's in holiday snap shots) and at other times are largely idiosyncratic (as when several people debate what's in one of Picasso's cubist paintings). Usually the pictorial space dominates awareness, and the painter Maurice Denis[2] deemed it necessary to write: 'Remember that a picture, before it is a war horse, a naked woman, or some anecdote, is essentially a flat surface covered with colors arranged in a certain order.' Indeed, the scientist should not forget this basic fact. On the other hand, the fact that one 'sees' a war horse or a naked woman in a mere arrangement of colours is surely something very remarkable indeed — this despite the fact that it is only the 'obvious' perception from the perspective of the naïve observer.

These initial observations on the topic of 'pictorial space' are perhaps sufficient to suggest that it is an elusive phenomenon, hardly amenable to scientific investigation at all, be it of an empirical or conceptual nature. Surprisingly, good progress has been made and one is able to study pictorial objects empirically in considerable quantitative detail. Moreover, the geometrical structure of pictorial space turns out to be rather tight and starts to be well understood in a formal sense. I will review recent progress in a summary fashion.

The Operational Definition of Pictorial Relief

How does one 'measure' pictorial space if pictorial space is only a figment of the mind? This is obviously a problem of experimental psychology. On closer examination of the question one finds that it perhaps needs to be reformulated in order to make sense. For in order to 'measure' something, that something has to exist independently of, and in the same sense as, the measuring tool. In order to measure the length of an object one places a yardstick next to that object and judges coincidences of extremities of the object with notches on the scale on the stick.

---

[1] Gombrich (1959).
[2] Denis (1890).

Similar observations hold for the use of gauges, sieves, and so forth. But pictorial objects are 'in the mind' of an observer. What one may record in physical space is only the overt behaviour of the observer (movement, colour, smell, and so on), though. For over a century (experimental psychology originated rather late in the history of science, with the establishment of Wundt's laboratory)[3] this obviated the possibility of true 'measurement' (in the sense of geometry: length, angle, spatial orientation, and so on) of pictorial objects.

But notice that it is easily possible to place a measuring tool in pictorial space! This is done routinely in photographs for scientific use where the geologist places his or her hammer next to the rock specimen to be recorded 'for comparison'. The comparison may take place in pictorial space as the observer (of the photograph) compares the size of the hammer (as a pictorial object) to the size of the rock (as another pictorial object). Thus a simple idea is this: place a gauge figure (a pictorial object) in the (equally pictorial) scene and let the observer judge the 'fit' (pictorially of course). With modern computer-based methods one may even control the gauge figure parametrically and place it under (manual) control of the observer.

Here is a simple instance of this general method.[4] As a gauge figure we use an ellipse with fixed major axis and parametrically variable minor axis and orientation in the picture plane. When superimposed over a part of the picture that appears as a surface ('pictorial relief') in pictorial space, such a gauge figure can appear in (at least) two quite different visual modes. (See Figure 10.1.) When the parameters are randomly chosen, chances are that the gauge figure looks like something that is superimposed on the pictorial scene but does not belong to it. Often it is seen as not in the pictorial scene, but rather as a mark on the physical picture. But for some narrow range of parameter values the ellipse appears in a very different mode. It is seen to *belong* to the pictorial scene, and appears as a circular mark on the surface of the pictorial object. When the observer manually controls the parameters it is as if the gauge figure suddenly attaches to the surface when the parameters are 'about right'. With a well-designed interface[5] the observer adjusts the gauge figure to lie on the surface effortlessly in just a few seconds. We routinely sample hundreds of points in half an hour or so.

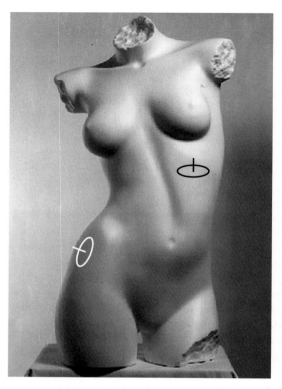

Figure 10.1 A stimulus (photograph of a torso) with superimposed ellipses (in the actual experiment only one ellipse is applied at any time). Notice the visual difference for the case of the ellipse that obviously 'fits' (the white one) and the one that clearly does not (the black one). Here we have added a normal sticking out of the surface of the disc in order to disambiguate the front/back confusion possible for a pure circular disc.

In the example given the essential task of the observer is the *judgement of coincidence*[6] of the plane of a circle in pictorial space with the tangent plane of an arbitrary surface in pictorial space at the location of the centre of the circle (say). The spatial resolution is limited by the diameter of the circle and has to be selected according to the roughness of the pictorial surface to be sampled. Where is the centre of the circle located? There are two components in the answer to this question. Pictorial space is a three-dimensional space. Two of its dimensions are easily identified with the dimensions of the surface of the picture as a physical object. The centre of the circle in this sense is determined by the centre of the ellipse relative to the pigmentation of the surface, and is under full experimental control (since everything is located in physical space). The third dimension of pictorial space is often denoted 'depth'. It is an elusive dimension that exists only in the mind of the observer. When the ellipse appears as a circle on the pictorial surface the centre of the circle is on the pictorial surface, and both have the same depth at that point. It is not under experimental control,

[3] Wilhelm Wundt opened the first formal psychology laboratory in 1879 at the University of Leipzig.

[4] Koenderink et al. (1992).

[5] Although I don't discuss the issue any further this is crucial. It is easy to implement an interface that will make the task an impossible one. I will not quote examples from the literature. A simple example of an 'impossible' interface is two turning knobs that control the Cartesian co-ordinates of a cursor in the plane. It takes long practice to use this for something as simple as writing your name. Interfaces like that have actually been used in psychophysics and have led to spurious beliefs on visual perception. There is no reason to expect such failures to be rare in the future.

[6] A judgement of coincidence avoids the (conceptual and practical) problems with 'magnitude estimation' so common in mainstream psychophysics. Judgements of coincidence or equality escape problems with 'qualia'. This is why 'colorimetry' is a science, whereas 'colour science' is not.

but simply 'happens'. The coincidence is a fact because it is seen to be the case. That the circle is on the surface is a fact in the observer's consciousness. In physical space the circle doesn't even exist and 'depth' is meaningless. The ellipse is at no particular 'depth' because it is not even located in pictorial space.

Since oblique circles appear as ellipses in projection, an obvious move is to interpret the ellipse parameters produced by the observer in terms of a 'depth gradient' or 'surface attitude' (say in terms of slant and tilt angles). Then the method yields a sample of the depth gradient at a point picked by the experimenter. Notice that this is an operational definition of the surface attitude and that the surface attitude cannot be said to exist in the absence of the probing.[7] Thus one should refrain from asking how closely the sample approximates the 'true' attitude. The only true attitude is the result of the sampling! In practice we sample at the vertices of some triangulation of the picture surface (unknown to the observer) and visit the vertices, one at a time, in random order. This yields a discrete sampling of a gradient field. (See Figure 10.2.) We proceed to find the best fitting surface to this field[8] (it is defined up to an arbitrary depth shift) and call this the 'pictorial relief'. Again, the relief is produced by the method, and it is a moot question as to whether it exists in the mind of the observer apart from the sampling. We treat this relief (a triangulated surface with a hundred to a thousand vertices) as the psychophysical 'response' of the observer to the 'stimulus' defined by the picture.[9]

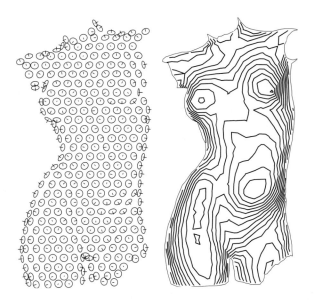

Figure 10.2 The result of a quarter of an hour's sampling (left), and the 'pictorial relief' (right). The pictorial relief is a surface in three-dimensional pictorial space and the 'response' to the 'stimulus' which is the image shown in the previous figure. Here the relief is plotted as a depth map via the pattern of equal depth contours.

This is just one example of a psychophysical procedure. Many different paradigmata are possible, but in this paper I will stick with the example. The important points are that a quantitative, geometrical data structure is obtained in response to a picture and that this data structure is both obtained and defined by the method.

Straightforward Observations, the Problem of Veridicality

With the psychophysical method in place, the next step is to obtain the pictorial relief for a number of observers and a variety of pictures. We find that observers find it a simple task to set the gauge figure on straightforward photographs of simple objects (say a piece of smoothly finished sculpture), but meet with increasing difficulties when either the rendering is deteriorated (e.g. cartoon style or even silhouette renderings instead of full-scale photographs) or the object is more complicated (e.g. a treetop). When results are compared we find that the responses of a single observer over time reproduce much better for straightforward photographs of simple things than for impoverished renderings or overly complicated objects. The concordance between responses of different observers follows the same pattern.[10]

In the simplest cases (straight photographs of moderately complex objects—we used human torsos or fairly smooth, abstract sculpture a lot—in frontal poses) observers tend to agree at least qualitatively. (See Figure 10.3.) However, the results fail to be 'veridical' in the naïve sense. (See Figure 10.4.) When we study the correlation of depth values at corresponding vertices for different observers, values of the coefficient of determination (R^2) are typically over 90 per cent, with values over 99 per cent not being particularly rare.

In performing such linear regression analyses we noticed that, though the correlations are typically impressive, the depth values are rarely identical. Although clearly linearly related, slopes up to five are routinely encountered. Such slopes represent linear scalings of depth, or 'depth magnifications'. These were already identified by the German sculptor Adolf Hildebrand[11] at the close of the nineteenth century. Hildebrand noticed that observers often confuse bas-relief renderings with sculpture in the round. Hildebrand's analysis was remarkable and surely far ahead of his time.

For a single observer such dilations and contractions of depth can reliably be induced via manipulation of the viewing conditions.[12] (See Figures 10.5 and 10.6.) Looking at the picture squarely or obliquely, with one or both eyes, with or without accommodation cue, and so forth, induces factors up to five. Such effects were

---

[7] Thus literature attempts to 'calibrate' the gauge figure attitudes via magnitude estimation of isolated gauge figures are abortive.

[8] This involves a check of integrability. We have never encountered fields with any significant curl, though.

[9] Koenderink et al. (1992).

[10] Koenderink et al. (1992, 1994, 1996a, 1996b).

[11] Hildebrand (1945).

[12] Koenderink et al. (1994).

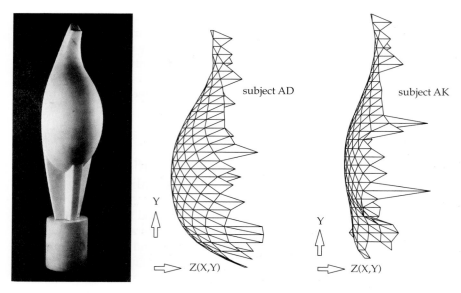

Figure 10.3 Pictorial reliefs for two observers of the same stimulus. The stimulus (left) is a photograph of the *Bird* sculpture by Brancusi from the Philadelphia Art Museum. Neither of the observers had seen the sculpture. Notice that the pictorial reliefs produced by observers AD and AK are quantitatively different (though qualitatively similar). Clearly at most one of the observers can be 'right'; most probably neither is.

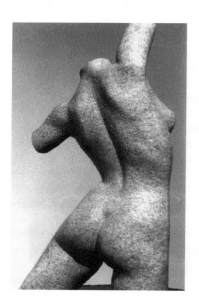
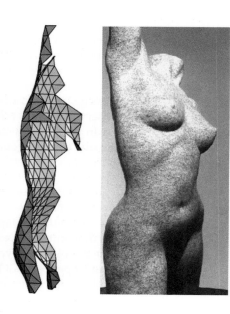

Figure 10.4 Pictorial relief (centre) of an observer for the stimulus (left) compared with a photograph of the fiducial object in another pose (right). The rendering of the pictorial relief and the choice of the pose in the photograph at the right are such that the contours of both should be identical in the case of true 'veridicality'. Clearly the response fails to be veridical in the naïve sense.

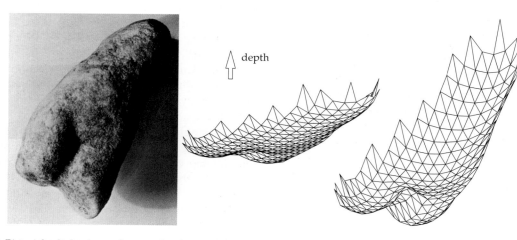

Figure 10.5 Pictorial reliefs of one observer for the same stimulus (left) for two different viewing conditions—binocular (centre) and synoptical (right). In either case both eyes are open; in the synoptical case the retinal images are identical, though. In the binocular viewing the regular disparity field induced through the lateral offset of the eyes is present; in the synoptical case it is not.

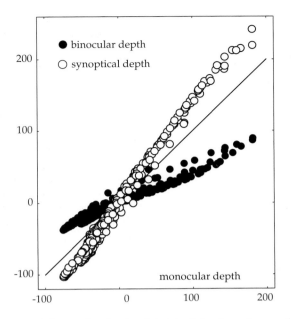

Figure 10.6 Scatterplots for the pictorial reliefs of one observer for the same stimulus (left of Figure 10.5) for three different viewing conditions (monocular, binocular, and synoptical). Clearly Hildebrand's interpretation works quite well in this case. Compared with monocular vision, binocular viewing lets the depth domain contract, whereas synoptical viewing lets it expand.

already (informally) known to Leonardo[13] and were exploited by the optical industry[14] from the second half of the nineteenth century up to (about) the 1950s to produce a variety of picture-viewing machines, many of which can be seen at science museums or found at flea markets.

Apart from these Hildebrand dilations (which are indeed ubiquitous) we also find examples of (occasionally very) low correlations between observers or for results obtained via only slightly different methods for a single observer.[15] (See Figure 10.7.) It took us a while before these cases were formally understood. They are almost without exception the result of 'mental changes of perspective' that reflect the 'beholder's share'. I discuss these interesting cases below.

One issue that comes up again and again (always in discussions with laymen, quite regularly in scientific debate) is that of 'veridicality'. As I have indicated in the introduction, the naïve notion of veridicality is void. The veridicality issue is a non-issue. There need not even *exist* a fiducial scene that might be used to judge the degree of veridicality or approximation of the pictorial to the physical structure. As we all know from the products that reach us from the Hollywood movie studios, pictorial things need not look at all what they are. Although a moot point, a closer analysis of the concept of veridicality in pictorial perception leads to some very valuable insights (see below).

13 Leonardo da Vinci (1989).
14 Carl Zeiss Jena (1907).
15 Koenderink et al. (2000, 2001).

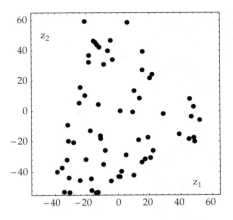

Figure 10.7 Scatterplot for the pictorial reliefs (the depth values $z_{1,2}$ at corresponding locations in the image) of one observer for the same stimulus (the photograph of the Brancusi *Bird* sculpture) in two (slightly) different tasks. Clearly Hildebrand's idea doesn't work at all here. The coefficient of determination (R^2) is not significantly different from zero. Mainstream psychophysics would either condemn the methods or throw a hundred observers at the task in order to obtain a 'significant' effect. This misses the point completely, as I will illustrate below.

Theoretical Considerations on the Concept of Veridicality

How many 'fiducial scenes' correspond to a given picture? Clearly infinitely many! Here are just two (extreme) examples: the picture surface with its distribution of pigments clearly 'explains' the scene. Then pictorial space is isomorphic with the physical picture. For the other example I move the 'pixels' out into the universe, using a random generator to determine how many light years. I replace the pixels with stars of exactly the right magnitude and colour to 'account for' the pixel. Then I have constructed a cosmic object (as different from the physical picture as I can imagine) that would yield the same retinal irradiance distribution and thus (if we are rational) must evoke the same 'pictorial scene'. Indeed, why not? Infinitely other scenes are imaginable that would yield the selfsame pictorial scene. One of these no doubt happens to be the exact physical scene at the time of exposure of the photograph. But is it in any way singled out among the infinite contenders? *Clearly no!*

Thus the concept of 'veridicality' is void in the sense that it doesn't apply to the observer. It applies to the larger set of the photographer and the observer, as well as the 'scientist' analysing the situation. Notice that any of these three persons may coincide — it makes no essential difference.

Let us assume that the picture is the result of some type of 'projection' (a first arbitrary assumption). In this paper I will concentrate on 'orthographic projection', but the exact projection really does not make much difference. Then each 'pixel' (point of the picture plane, if you want) stands for a 'visual ray' in the Euclidean sense.[16] The point corresponding to the pixel in the 'fiducial scene' has to lie on a line through the pixel and extending

16 Euclid's 'Optics' dates from c. 300 BC.

in the depth dimension. If I shift the points along their lines I create different scenes that are all equivalent in the sense that they would yield the same picture. A smooth pictorial surface may be assumed to correspond to a smooth surface in the fiducial scene (another arbitrary assumption). Then I may transform scenes through smooth isomorphism that conserves visual rays and thus obtain all equivalent scenes that explain the picture and satisfy the assumptions. I may add other assumptions — for example, radiometric properties. For instance I may assume that the pixel intensities are due to surfaces with certain scattering properties (Lambertian is popular in scientific circles, even though such surfaces don't exist in nature) irradiated in certain ways (collimated beams are popular assumptions), and so forth. Then the set of equivalent scenes shrinks in the geometrical sense, but expands in the physical sense (it now includes descriptions of the sources, for instance). By now it should be clear what I'm hinting at, namely that infinitely many scenes that 'explain' the picture exist and that the exact nature of this equivalence set depends critically upon a number of arbitrary assumptions.

Observers live in the physical world. Their heads are in the world and — equally important — the world is in their heads. It is only in this way that they can conceivably exist as efficacious agents; the causal nexus of the physical world is 'mirrored' in their mental makeup and physiological implementation. The only logical alternative (which cannot be considered seriously in the sciences, though) is predestination. Thus observers may be said to possess a working knowledge of 'ecological optics'. They blindly ascribe (unconsciously) to a number of prior hypotheses that reflect their competence in the generic structure of the causal nexus. Among these is the hypothesis of projection (visual rays) and a mixed bag of fragments of 'laws' (irradiance depends monotonically upon surface obliquity with respect to the average flux vector) and 'generic assumptions' (things are coherent regions of space bounded by smooth surfaces; most surfaces scatter in a Lambertian fashion, are opaque, and so forth). The prior assumptions are often denoted 'cues', a concept originated by the remarkable Bishop Berkeley.[17]

The theory of 'visual cues' has always played a role in experimental psychology and was expanded by visionaries such as Gibson.[18] However, the formal analysis of a fair number of cues is due to the relatively new discipline of computer vision. As a result, at present we understand the physics and mathematical structure of a number of cues under rather extreme simplifying assumptions.[19] It will probably take many decades of work before a more

intricate understanding of at least the major cues important in daily life vision has been arrived at. Thus the theories as developed today are quite insufficient to allow a more or less precise demarcation of the equivalence set of scenes for any given photograph. One needs a break.

A simple reasoning that at least gets us in the right ball-park is the following. First of all one ascribes to the projection hypothesis. That is to say, each pixel is due to some source (typically a scattering surface element) located upon the visual ray defined by that pixel. The second assumption is inspired by an analysis of a number of well-known cues (affine shape from motion, shape from shading, and so forth). It appears to be the case that the property of planarity of uniform surfaces in uniform compartments of the environment is revealed by virtually all known cues. Take the example of 'shading': if you encounter a uniform patch in the picture and it appears that this patch belongs to a surface, then that patch is probably planar, though its attitude and albedo must remain unknown. Any non-uniform pigmentation of the surface and any deviation from planarity are likely to generate non-uniformities in the picture. Certainty cannot be reached, but the hypothesis that this cue conserves planarity under variations of parameters (spatial attitude, direction of illumination) is a rather dependable one. A similar reasoning applies to many of the cues (dozens of them).

If the projection hypothesis and the conservation of planarity are accepted, the set of equivalent scenes is settled. Starting from some possible scene the others can be generated through application of the group of transformations that conserve both planarity and the visual rays. The elements of this 'group of ambiguities' are affinities that conserve a pencil of parallel lines[20] (the lines running purely in depth). The free parameters of this group are a depth shift, a depth magnification, and two components of an arbitrary shear that conserves planes containing a visual ray; thus it is a four-parameter group. If you include arbitrary similarities of the picture plane (two parameters for the translation, one for the rotation, one for the magnification), you obtain an eight-parameter group of 'pictorial space similarities'. It is the analogue of the seven-parameter group of Euclidean similarities of physical space.

Factoring Out the Beholder's Share

With the construction of the group of ambiguities of pictorial space we have effectively constructed the 'beholder's share'. The ambiguities can be freely applied by the observers without violation of the evidence

[17] Berkeley (1975). Berkeley insists that the cues are *arbitrary* (when you look someone in the face and suddenly notice a spectral red-shift, why should you 'see' shame?). This seems at variance with our present 'understanding' of many of the 'depth cues'. Of course the theory of evolution has provided the solution to this. Both Berkeley's arbitrary nature and our 'understanding' of cues are easily accommodated.

[18] Gibson (1950).

[19] For instance, see Koenderink and van Doorn (1980, 1997); Horn and Brooks (1989); Faugeras (1993).

[20] I stick to orthographic (or pseudoperspective) projection in this paper. Although easily generalised, I think it is actually a fit choice in most cases of pictorial perception. Observers typically don't look at pictures monocularly, with the eye located at the exact perspective centre. They somehow relate to the picture plane, even when viewing it obliquely.

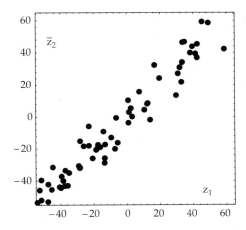

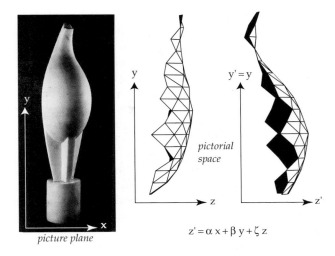

$$z' = \alpha x + \beta y + \zeta z$$

picture plane

Figure 10.8 A multiple regression for the pictorial reliefs of one observer for the same stimulus in two (slightly) different tasks. These are the same data as shown in Figure 10.7, but the 'depth' values $\bar{z}_2$ have now been corrected with a linear combination of the $\{x, y\}$ co-ordinates of the points. Although Hildebrand's idea does not work here, the multiple regression reveals that the responses are almost identical after all! The coefficient of determination (R^2) rises from not significantly different from zero to over 90 per cent. When the responses are compared without factoring out the beholder's share they appear as virtually unrelated. Mainstream psychophysics would have rejected these data out of hand.

Figure 10.9 A stimulus and two pictorial reliefs (same observer) obtained via two (slightly) different tasks. Notice how the two reliefs look 'rotated' with respect to each other. A multilinear regression involving (apart from the depth) the Cartesian co-ordinates in the picture plane reveals that the two responses are virtually identical (coefficient of determination 99 per cent) except for a movement of the mental eye (ambiguity transform). The coefficient ζ would specify a Hildebrand scaling, while (α, β) specify a 'movement of the mental eye'.

provided by the picture. Any transformation must necessarily lead to a pictorial scene that provides as valid an interpretation of the picture as any other. What this means is that the response of the observer in any psychophysical experiment is only causally dependent on the stimulus in so far as we factor out the ambiguities. One has to understand the 'response' modulo *arbitrary ambiguities*.

This is a very important insight that allows us to make sense of many cases in which we have found very low correlations between observers or between responses of a single observer in different tasks. What should be done is to compare the responses modulo arbitrary ambiguities. Thus one applies a Procrustean method in which one or both of the entities to be compared is transformed until a 'best fit' is obtained. It is only this best fit that has any intrinsic relevance. Because the ambiguities are linear transformations the comparison of responses takes the form of a multilinear regression in which figure not only the depth values, but in addition the Cartesian co-ordinates of the vertices of the triangulation in the picture plane.

We have found repeatedly that the results of such analysis are little less than astounding. (See Figure 10.8.) In many cases we find that a straight regression suggests that the responses are quite independent and thus idiosyncratic (R^2 not significantly different from zero), whereas the multilinear regression reveals a very high correlation (R^2 over 90 per cent). In such cases the responses are actually *identical* in so far as they causally depend upon the stimulus (the picture), although they appear quite different at first blush. The ambiguities then appear as the 'beholder's share'. Apparently the beholder's share is indeed fully idiosyncratic, but yet rather limited. It cannot be anything but a member

of the group of ambiguities. Because the beholder's share involves arbitrary similarities of pictorial space, it may be understood as arbitrary changes of 'mental perspective' or the execution of 'movements of the mental eye'. (See Figure 10.9.)

The study of such changes of mental perspective is pure psychology and has to depend upon the *semantics* of the picture, since the cue dependency ('syntactical structure') has already been exhausted in the construction of the geometrical (meaningless) part of pictorial space.

The Structure of Pictorial Space

If one understands (as is indeed reasonable) the group of ambiguities as the group of similarities of pictorial space, then this group may be said to define the geometrical structure of pictorial space. This is the view of 'geometry' as pioneered by Felix Klein in the nineteenth century.[21] Indeed, it can be shown that the ambiguity group leads to one (of the twenty seven) 'Cayley–Klein' geometries of dimension three. It is the space with two Euclidean and one 'isotropic' dimension. This space has a number of remarkable properties that set pictorial space apart from physical space.

From the perspective of projective geometry the geometry induced by the ambiguity group is defined by the fact that the 'absolute quadric' (which we locate in the plane at infinity) is degenerated into an intersecting line pair. The metric is the degenerated form $ds^2 = dx^2 + dy^2$. Factoring this expression we find indeed that the isotropic entities satisfy $(dx - i\,dy)(dx + i\,dy) = 0$. These are families of parallel planes $dx = \pm i\,dy$ and a family

---

[21] Klein (1871, 1872).

of parallel lines $dx = dy = 0$. The lines meet the plane at infinity in a point and each family of planes meets the plane at infinity in a line. Thus the (degenerated) absolute quadric in the plane at infinity is composed of two intersecting lines. Because the absolute quadric is self-dual the metric of the space is dual. We obtain a full metric duality of lines and planes. The 'full isotropic lines' (the lines that intersect both lines of the absolute quadric) are all parallel to the z-axis: that is to say, they are visual directions.

The metric is possibly best understood[22] as a limiting case of the Minkowski metric $ds^2 = dx^2 + dy^2 - dz^2/\gamma^2$ for $\gamma \to \infty$. Then the visual rays are understood as degenerated 'light cones' and the spatial order is immediately apparent. On a single visual ray a depth order exists, but points on different visual rays cannot be compared with respect to depth. There is always a movement to be found that will bring them in frontoparallel position.

The general ambiguity can be parametrised as

$$\begin{pmatrix} x' \\ y' \\ z' \\ 1 \end{pmatrix} = \begin{pmatrix} h_1 & -h_2 & 0 & t_1 \\ h_2 & h_1 & 0 & t_2 \\ c_1 & c_2 & f & t_3 \\ 0 & 0 & 0 & 1 \end{pmatrix} \begin{pmatrix} x \\ y \\ z \\ 1 \end{pmatrix}$$

It depends upon eight parameters. The most important subgroup ($h_1 = 1, h_2 = t_1 = t_2 = 0$) describes our results; for $c_1 = c_2 = 0$ we obtain the two-parameter subgroup of Hildebrand transformations.

When you treat the space modulo the full isotropic lines you simply obtain the picture plane, for each 'pixel' corresponds to a visual ray, and thus to a full isotropic line. Thus each spatial configuration has a natural 'trace' (its footprint by the visual rays) in the picture plane. The picture plane is a Euclidean plane with the metric $ds^2 = dx^2 + dy^2$. All metrical relations in pictorial space thus are identical to these of their traces in the picture plane. Notice that planes that neither intersect nor coincide are possible; they are parallel. Duality then implies the existence of 'parallel points'. Such points lie on a single visual ray (Bishop Berkeley would have held that they cannot be distinguished optically)[23] and are separated by a gap that extends purely in 'depth'. We may use the gap-width as a makeshift 'distance'. If we redefine distance subtended by two points as either the distance of their traces (the generic case) or the gap (in case they happen to be parallel), we obtain a quantity that is invariant under actions of the subgroup with $h_1^2 + h_2^2 = f = 1$. This is the group of proper 'movements'. The general group includes 'similarities', which are seen to depend on two parameters, $\sqrt{(h_1^2 + h_2^2)}$ and $f/\sqrt{(h_1^2 + h_2^2)}$. The first one scales distances, the second one scales slants (in depth). This is different from Euclidean space where similarities depend upon a single scaling factor.

It is of some interest to consider the 'rotations' of pictorial space. There exist a few types, some appearing as rotations, some as translations, and some as identities

in the traces in the picture plane. The interesting ones are the latter (they correspond to movements of the mental eye). For such 'rotations' the angle metric is parabolic (like the distance metric on a line). Thus the 'angle' takes values between plus and minus infinity and is not periodic (as it is in the picture plane). Thus you cannot turn around in pictorial space. Informally, if you look at a photograph of a person taken frontally you will never be able to see the back of the head, no matter what mental eye movements you may attempt. The reason is obvious: What is not 'in the picture' can never be seen because it fails to exist. Thus it is immediately obvious that pictorial space has to be a non-Euclidean space.[24]

The geometry of pictorial space is quite different from that of Euclidean space, but every bit as rich. Good introductions to the 'geometry of isotropic space' are available.[25]

The differential geometry of pictorial space is especially important, because the differential invariants define the 'pictorial shape features'. The lowest order of interest is the second; it deals with the 'curvature' properties. The familiar 'gradient space' $\{z_x, z_y\}$ from machine vision can be shown to be isometric with the isotropic space analogue of the 'spherical image' of Euclidean differential geometry. (As defined via the orientations of tangent planes, the 'normals' in pictorial space are all parallel to the viewing direction!) The spherical image can be used to find the second order invariants. They turn out to be simply the eigensystem of the Hessian of the depth. Thus the 'Gaussian curvature' is $z_{xx}z_{yy} - z_{xy}^2$ and the mean curvature $(z_{xx} + z_{yy})/2$. Here the Gaussian curvature is defined as the magnification (Jacobean) of the spherical image: the 'intrinsic' curvature is identically zero since the metric is flat and all normals parallel to the visual direction. The formalism of isotropic differential geometry is much simpler (all formulas are shorter) than that of Euclidean differential geometry,[26] although almost all Euclidean properties have close analogues in the geometry of isotropic space.

The understanding of the structure of pictorial space thus allows one to predict the elements of 'pictorial shape'. Such predictions are immediately subject to psychophysical investigation. On a very shallow (and thus general) base we have erected a powerful theory of pictorial space.

Concluding Remarks

In this paper I have reviewed my current understanding of pictorial perception and the structure of 'pictorial space'. I cannot speak of a consensus, for the psychophysical literature is unfortunately very confused, with

---

[22] Yaglom (1979).
[23] Berkeley (1975).

[24] This also implies that many analyses current in mainstream psychophysics make no sense. For instance, Euclidean normals on a pictorial surface are meaningless, yet often used. In pictorial space all normals are isotropic lines (parallel to the viewing direction).
[25] Strubecker (1941, 1942, 1943, 1945); Sachs (1990).
[26] Koenderink (1990).

many misunderstandings, invalid empirical claims, and wild speculations. The reasons for this unfortunate state of affairs are largely conceptual — an undue preoccupation with the (empty) naïve notion of veridicality — but also of a technical nature (the standard methodologies don't allow one to obtain geometrical data from the observer's responses). It seems to me that the way I have outlined the matter here cannot fail to be part of the framework of any viable theory of pictorial space. The reader should be aware that I'm a loner in psychophysics, though.

Several changes in viewpoint (as compared with the mainstream of the psychophysical literature) have enabled us to make some progress.

1. The insight that the naïve notion of veridicality is void has been absolutely crucial. It has many important consequences: for instance, many of the standard analyses can be skipped (because seen to be senseless), and more stress is therefore laid upon an analysis of the intrinsic structure of the responses (as distinct from the extrinsic structure which depends upon comparison with some fiducial scene).

2. The design of psychophysical methods that allow one to obtain large datasets in a short period has been essential. The responses need to be of a geometrical nature (e.g. large triangulations), thus one needs to sample the equivalent of at least a thousand numbers per hour.

3. The psychophysical methods should imply no 'stimulus reduction' at all since this prohibits the use of pictures that evoke structured pictorial spaces to begin with. Virtually all examples of studies based on 'stimulus reduction' have yielded sterile results.

4. Response reduction is necessary in order to obtain objective and reproducible results; mere judgements of coincidence or equality are strongly preferred. This may be the only way to skip the problem of 'qualia' and yet come up with something useful.

5. The insight that 'responses' should be taken modulo the beholder's share has been crucial. In my view many published results are essentially rendered worthless because of an ignorance of this fact. It seems very likely to me that many 'problems' in psychophysics might well disappear when responses are projected upon suitable spaces. Unfortunately this topic has received no attention at all in the literature.

The theory of the structure of pictorial space is closely related to a general understanding of the intrinsic ambiguity of pictures and to an understanding of observers as efficacious agents due to a working knowledge of ecological optics obtained through interaction with the physical world. The theory has been derived from psychophysical data and indeed is necessary to analyse the data (by factoring out the fully idiosyncratic movements of the mental eye). Because the psychophysical data are extensive and rich and do not depend upon unrealistically simplified stimuli (we mostly use photographs of real scenes, and hardly ever rely on computer graphics),

the theory is solidly grounded in fact. The theory allows us to predict a variety of 'pictorial shape features' that are up to empirical verification; thus it goes much beyond the immediate data.

The theory and the psychophysical methods subtend a principled field of endeavour, rare in the experimental psychology of perception. Fields such as colorimetry are in a similar position in that they involve large amounts of coherent quantitative data and theory with excellent predictive power. However, such fields only 'work' because most of the action is at the very periphery of the brain. For instance, in the case of colorimetry everything is essentially settled at the level of the absorption of photons in the retinal receptor's outer segments. Such fields depend essentially on simple physical processes rather than on higher brain processes, rendering them perhaps somewhat trivial from a conceptual perspective (though very useful from a pragmatic viewpoint). This is quite different in the case of pictorial perception. Here most of the action is at the higher levels of the visual system, which remain largely virgin territory to the neurophysiologist. Of course the topic of 'pictorial space' is currently far outside the reach of neurophysiology; it can only be approached through psychophysical methods. It is my conviction that it will remain this way because there are no neurophysiological methods that address the structure of consciousness (pictorial space is a thread of consciousness) in any direct manner. Pictorial space is by its very nature an 'intentional' entity[27] that is not accessible to physiological methods.[28]

Note

Andrea van Doorn helped me prepare this paper. Andrea van Doorn, Astrid Kappers, Joseph Lappin, and James Todd participated in many of the psychophysical experiments on which this paper is based, and I owe much to our conversations.

References

BERKELEY, G. B. (1975), *An Essay towards a New Theory of Vision* (1st edn. 1709, Dublin), in *Philosophical Works*, ed. M. R. Ayers (London).

BRENTANO, F. VON (1874), *Die Psychologie vom empirischen Standpunkte* (Leipzig).

[27] Brentano (1874).
[28] Suppose one found a fairy world 'lit up' in a brain scanner and that it resembled people's 'pictorial space' for the given stimulus. Would one have the ultimate 'explanation' of our pictorial space? No! The fairy world might conceivably be produced (given suitable progress in machine vision) directly from the stimulus, avoiding any brain activity. Brain activity can only destroy structure causally related to the stimulus. Some intermediate brain activity (as long as it retained the relevant structure) would do no harm, though it would be irrelevant. Such a feat would gain a Nobel prize in medicine, but would be essentially meaningless from my point of view.

CARL ZEISS JENA (1907), *Instrument zum beidäugigen Betrachten von Gemälden u.ggl.*, Kaiserliches Patentamt, Patentschrift Nr. 194480, Klasse 42h, Gruppe 34.

DENIS, M. (1976; orig. 1890), in R. Goldwater and M. Treves (eds.), *Artists on Art, from the 14th to the 20th Century* (London): 380.

EUCLID (1959), *Euclide: L'optique et la catoptrique*, trans. P. V. Eecke (Paris).

FAUGERAS, O. D. (1993), *Three-dimensional Computer Vision: A Geometric Viewpoint* (Cambridge, MA, and London).

GIBSON, J. J. (1950), *The Perception of the Visual World* (Boston).

GOMBRICH, E. H. (1959), *Art and Illusion*, part III, 'The beholder's share' (London).

HILDEBRAND, A. (1945), *The Problem of Form in Painting and Sculpture*, trans. M. Meyer and R. M. Ogden (New York). Originally published in 1893 as *Das Problem der Form in der bildenden Kunst* (Strassburg).

HORN, B. K. P. and BROOKS, M. J. (1989), *Shape from Shading* (Cambridge, MA).

KLEIN, F. (1871), 'Über die sogenannte nichteuklidische Geometrie', *Mathematische Annalen Bd.* 6: 112–45.

KLEIN, F. (1872), *Vergleichende Betrachtungen über neue geometrische Forschungen*, Programm zu Entritt in die philosophische Fakultät und den Senat der Universität zu Erlangen [the 'Erlangen Program'] (Erlangen).

KOENDERINK, J. J. (1990), *Solid Shape* (Cambridge, MA).

KOENDERINK, J. J. and DOORN, A. J. VAN (1980), 'Photometric invariants related to solid shape', *Optica Acta* 27: 981–96.

KOENDERINK, J. J. and DOORN, A. J. VAN (1997), 'The generic bilinear calibration–estimation problem', *International Journal of Computer Vision* 23: 217–34.

KOENDERINK, J. J., DOORN, A. J. VAN and KAPPERS, A. M. L. (1992), 'Surface perception in pictures', *Perception & Psychophysics* 52: 487–96.

KOENDERINK, J. J., DOORN, A. J. VAN and KAPPERS, A. M. L. (1994), 'On so-called paradoxical monocular stereoscopy', *Perception* 23: 583–94.

KOENDERINK, J. J., DOORN, A. J. VAN, CHRISTOU, C. and LAPPIN, J. S. (1996a), 'Shape constancy in pictorial relief', *Perception* 25: 155–64.

KOENDERINK, J. J., DOORN, A. J. VAN, CHRISTOU, C. and LAPPIN, J. S. (1996b), 'Perturbation study of shading in pictures', *Perception* 25: 1009–26.

KOENDERINK, J. J., DOORN, A. J. VAN, KAPPERS, A. M. L. and TODD, J. T. (2000), 'Directing the mental eye in pictorial perception', in B. E. Rogowitz and T. N. Pappas (eds.), *Proceedings of SPIE: Human Vision and Electronic Imaging* 3959: 2–13.

KOENDERINK, J. J., DOORN, A. J. VAN, KAPPERS, A. M. L. and TODD, J. T. (2001), 'Ambiguity and the "mental eye" in pictorial relief', *Perception* 30: 431–48.

LEONARDO DA VINCI (1989), *Leonardo on Painting*, trans. M. Kemp and M. Walker; ed. M. Kemp (New Haven, CT).

SACHS, H. (1990), *Isotrope Geometrie des Raumes* (Braunschweig/Wiesbaden).

STRUBECKER, K. (1941), 'Differentialgeometrie des isotropen Raumes I', *Sitzungsberichte der Akademie der Wissenschaften Wien* 150: 1–43.

STRUBECKER, K. (1942), 'Differentialgeometrie des isotropen Raumes II', *Math. Z.* 47: 743–77.

STRUBECKER, K. (1943), 'Differentialgeometrie des isotropen Raumes III', *Math. Z.* 48: 369–427.

STRUBECKER, K. (1945), 'Differentialgeometrie des isotropen Raumes IV', *Math. Z.* 50: 1–92.

YAGLOM, I. M. (1979), *A Simple non-Euclidian Geometry and its Physical Basis: an Elementary Account of Galilean Geometry and the Galilean Principle of Relativity*, trans. A. Shenitzer, ed. assist. B. Gordon (New York). pp. 1009–26.

CHAPTER 11

The Potential for Image Analysis in Numismatics

Christopher J. Howgego

The application of new imaging techniques to numismatics has not progressed beyond the use of digital images in connection with databases, and for web and conventional publication. Such applications are of real importance but lie well within the scope of what can be done already, and hence are not of interest here. It is unlikely that image enhancement will be of great significance, except, perhaps, in microscopic metallurgy, because numismatists can normally see and reproduce what is necessary for the purposes of study. Coins were mass-produced: if one specimen is hard to read, it will usually be more useful to locate a better specimen than to enhance the image of the poor one. Where there is spectacular scope for progress is in image comparison, but no real start has been made in this area. This paper is therefore about significant potential, rather than current developments.

The systematic study of coinage at the level of the individually engraved die (alongside the scientific study of coin hoards) revolutionised numismatics, but the laborious nature of such work has severely limited its application.[1] Die studies are crucial not only for attribution and chronology, but also for quantification. There remains a great deal of evidence to be exploited, particularly for economic history. This paper sets out the potential for image analysis in this area, and also the principal technical challenges, with a view to stimulating discussion about the best way forward. The technical difficulties are significant, but the problems are interesting and solutions may have wider applications.

The development of the methodology of the die study and its application from the 1870s for mint attribution were themselves linked to a major revolution in imaging, namely the advent of photography.[2] It is a thesis of this paper that the new revolution in imaging has the potential to facilitate another quantum leap in the subject.

The Die Study

The die study is a simple but very powerful method. Most coins in antiquity were manufactured by striking a disk of metal ('blank' or 'flan') between two

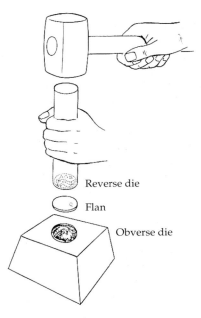

Figure 11.1 Schematic diagram of coin striking.

hand-engraved punches ('dies') (Figure 11.1). Few actual dies survive, but it is possible to gain a reasonably full knowledge of the original dies through visual comparison of the coins struck from them. Individual dies may well have struck 20,000 or 30,000 coins, and it is normal for each known die to be represented by a number of surviving coins. Statistical techniques allow us to estimate the original number of dies from the degree of duplication among the surviving examples. Quantification is to some degree controversial, not least because the range of the number of coins struck from a die can only be estimated from comparative studies and experimental archaeology. Even without this difficulty, calculations often produce results with wide margins of error. But no mint records survive from the ancient world, and the original number of dies is our best guide to the absolute size of an issue of coins.[3] The topic is important both for financial history (since one of the main reasons for striking coins in antiquity was to enable states to make payments) and for economic history (where the quantity of coins in circulation informs our models of money use).[4]

---

[1] Dies: Kraay (1976), xxi–xxiii, 11–19; Göbl (1978), vol. I, 220–2; Crawford (1983), 207–14; Mørkholm (1991), 12–19. Hoards: Kraay (1976), xxiii–xxv; Crawford (1983), 190–207; Crawford (1990).
[2] Mørkholm (1982).

[3] De Callataÿ (1995).
[4] For examples of the use of quantification to address financial history and money use respectively, see Crawford (1974); Walker (1988). For the broad issues and an analysis of the problems involved, see Howgego (1990, 1992).

Christopher J. Howgego

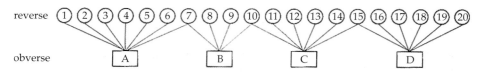

Figure 11.2 Idealised diagram of die links.

The lower die ('obverse'), protected in an anvil, and the upper die ('reverse'), in the form of a hand-held punch, wore out at different rates, and were replaced as necessary. Thus it is possible to identify sequences of dies linked together by studying the coins struck from them (Figure 11.2). Coins produced from a linked sequence may with reasonable certainty be attributed to broadly the same context (mint and date), so that the die study marked a radical improvement on earlier subjective attributions by style alone.

Our ability to utilise die studies for attribution and quantification has been limited by the sheer quantity of comparisons that have to be made for large issues of coinage. In principle, within each issue (defined for this purpose as a group of coins with the same images and inscriptions, and of the same denomination), each coin must be compared with every other coin. In practice, large issues may be broken down into smaller groups by detailed considerations (for example, the spacing of the inscription) so that there is no possibility of die links between groups (and hence no need to make comparisons). But the procedure remains a daunting prospect for very large coinages. Thus the coinages of Athens and Rome at the heights of their respective empires were so huge that few substantial die-studies have been attempted.[5] To give an idea of scale only, the principal coinages in gold and silver in circulation in the Hellenistic world around 290 BC have been estimated to have been struck from 65,000 drachma-equivalent dies,[6] and the Roman Republican mint is believed to have used up to 2400 dies for silver denarii in a single year.[7] It is perhaps symptomatic that the two largest die-studies of particular coinages under the Roman empire remain unpublished.[8] In general the application of the methodology has been patchy: die studies have often been undertaken where needed for attribution, but rarely for quantification alone.

Crawford made a pioneering attempt to quantify the coinage of the Roman Republic by extrapolating from a few small die-studies, using the relative frequency of issues in hoards as an index of their relative size.[9] Duncan-Jones applied a similar technique to the imperial coinage of the Roman principate.[10] Such attempts are to be applauded, but introduce further and very significant margins of error. Moreover, although hoards can give an idea of relative size, die studies provide the only significant approach to absolute quantification. No one would seriously doubt that it would be better to have both full die-studies and indices of frequency in hoards, so that these two approaches to quantification, each with their own problems, may be used to control each other.

The scale of the material presents a problem for taxonomy and attribution, as well as for quantification. The 'Berlin Corpus', the initiative of Mommsen and the great Swiss numismatist Imhoof-Blumer to create a classification of all Greek coins, has suffered from slow progress following the 'scientifically correct' but burdensome decision to work at the level of the die (the crucial debate took place in the 1890s and early 1900s).[11] Thus the sheer quantity of comparisons to be made in die studies is an obstacle to the advancement of the subject in a number of fundamental areas. It is here that computerised comparison of digital images has great potential.

Problems

Such a project will not be easy because the images to be compared have been subject to many distortions, at the time of striking (I), through subsequent wear and damage (II), and at the time of image-capture (III). If computerised comparison is to be effective, it will be necessary to find a way of identifying and discounting such distortions.

Distortions at Time of Striking

The fact that coins were struck by hand introduces a range of distortions in the images to be compared.

1   The penetration and angle of strike varies. The human eye (at least) can cope with this easily in most cases, but some instances may be very misleading. Modern experimental striking has produced unexpected variations between products of the same die (Figure 11.3).[12] The problem is at its most extreme in those archaic coinages which were struck on one side with an 'incuse punch' (Figure 11.4). The degree to which the punch was driven into the metal, and the angle of entry, produced enormous variations in the shape of the indentation on the coin (marks within the punchmark may be a better guide to die identity).

---

[5] Kraay (1956) produced a classic die-study of a Roman imperial coinage. See below, n. 8.

[6] The estimate of actual number of dies was lower: the use of 'drachma-equivalent dies' is a convention to express comparative value. For example, a tetradrachm (four drachma) coin die is counted as four drachma-equivalent dies. See de Callataÿ et al. (1993), 43, for 290 BC, and annexe I (p. 97) for later issues; de Callataÿ (1997).

[7] Crawford (1974).

[8] Bland (1991); Doyen (1989).

[9] Crawford (1974). The Lohe hoard (eighteenth-century Sweden) provided the classic demonstration of a correlation between frequency in hoards and mint output: see Volk (1987).

[10] Duncan-Jones (1994).

[11] Mørkholm (1982), 15–16; Kaenel (1991).

[12] Beer Tobey and Tobey (1993).

Figure 11.3 Two products of the same die from a modern experimental striking. Note that the neck of the turtle appears to vary in length.

Figure 11.4 Coins from the same reverse punch (R91), demonstrating variability according to angle and penetration of strike. From an unpublished die study, courtesy of H. S. Kim.

Figure 11.5 Coin struck off-centre. Plaster cast of a silver coin of Augustus.

Figure 11.6 Double-struck coin, with the letters SC and wreath off-set. Plaster cast of a bronze coin of Augustus.

Figure 11.7 Die break across the back of a turtle. Plaster cast of a coin from Aegina.

Figure 11.8 Ear of barley overstruck on a winged horse (Pegasus), oriented to show the undertype, and a Pegasus for comparison. Plaster casts of coins from Metapontum and Corinth.

The only consolation is that die studies of such material are by far the hardest to perform in all ancient numismatics. The positioning of the image on the flan varies, and it is not uncommon for coins to be struck sufficiently off centre for part of the image to be missing from the coin altogether (Figure 11.5).

2 The flan may be struck more than once by the die and the two impressions may not be precisely aligned, resulting in a ghosting (Figure 11.9 below) or plainly disjointed effect (Figure 11.6) — 'double struck'.

3 The die itself may be damaged in the course of use, and the 'die break' may be progressive (Figure 11.7). The damage to the die shows up as an additional feature in relief on the coin (so that a crack in the die results in a ridge on the coin). This is not all bad news: the numismatist knows that two similar coins with consistent die-breaks are virtually certain to be from the same die. Presumably a computer could also be trained to use this short cut. Die breaks may also be useful in providing the chronological direction of a die sequence: a coin with a particular die-break must have been struck later than a coin from the same die without the die break.

4 A further complication, which may also be beneficial, is that, sometimes, existing coins were used as flans (resulting in an 'overstrike'). The image on the original coin may show through and thus 'distort' the superimposed image (Figure 11.8). Such distortions must be discounted for the purpose of die comparison, but overstrikes also provide vital information for the numismatist in the form of a cast iron stratigraphy. The coin used as a flan must be earlier than

the coin struck over it. It can be extremely difficult to recognise the undertype from the traces left, and computerised imaging might facilitate this by 'lifting' one image off the other, perhaps in much the same way that two superimposed texts may be separated.

Subsequent Distortions

Once in circulation individual specimens may be damaged in a variety of ways. They may be holed (Figure 11.9) or cut (Figure 11.10), or deliberately stamped with punches bearing images and/or inscriptions ('countermarks') (Figure 11.11). They may also be corroded (either adding to, or subtracting from, the surface), and surface colour may be varied. Surface discoloration may be a significant problem for coins known only from directly taken images (photographic or digital), which is why numismatists have often preferred to work from plaster casts. But by far the most significant of these problems will be mundane coin wear — almost all coins are worn to some degree (Figure 11.10). Somehow it will be necessary to model wear, possibly drawing on the packages used to model landscape erosion.

Image Capture

The coins may be available to be imaged under fixed conditions, but they are in many locations, and the numismatist also has to cope with plaster casts, conventional photographs, and digital images. Perhaps 40 per cent of the material is known only from photographs in trade sale catalogues, and increasingly from commercial digital images on the web. So there can be no ducking problems arising out of the variability of image capture.

Figure 11.9 Two coins from the same die, one holed, the other slightly double-struck (notice the A of the inscription). Plaster casts of coins from Abdera.

Figure 11.10 Two coins from the same die, both cut and one very worn. Plaster casts of coins from Abdera.

Figure 11.11 Two countermarks, of an imperial head and a monogram, to the right of the imperial bust. Plaster cast of a Roman coin from Laodicea.

Shadows are an obvious problem for 2D images which have been taken under variable lighting conditions. The direction, nature, intensity, and number of the light source(s) may all vary. Controlled stereo imaging or 3D imaging might provide an answer for new images, but cannot help with the substantial body of existing images. Presumably similar problems of shadow are addressed in the analysis of aerial and satellite photographs.

Solutions

To make an impact we need not only a suitable technology for image capture, but also a powerful facility for automated image comparison. This need not be perfect: it would still be helpful to be given all certain and probable die-links (the level of probability needs consideration). One could then settle for working within defined margins of error (for the purposes of quantification), or resolve 'difficult cases' by visual inspection where possible. Even so, a major challenge will be to define key variables, and to model forms and levels of variability displayed by products of the same die, and to differentiate that from variability between products of different dies.

It is possible to work at the level of only 2D images: the human eye and brain can perform a perfectly satisfactory die-study from photographs alone. Here fingerprint technology might have interesting lessons to teach, but in numismatics no progress has been made beyond visual comparison of images superimposed on film or screen.

However, to reduce 3D objects (albeit with shallow relief) to 2D may be to throw away valuable information which could help with some of the distortions outlined above. For example, one might model a die from a surviving coin and then fit it to another coin. This might help to resolve problems resulting from angle of striking and degree of penetration, and from striking off-centre. It might also provide an approach to coin wear. Since wear can only detract from relief, and not add to it, a 'die' fitted to a coin from that die cannot overlap it (the wear would show up as a gap between 'die' and coin). 3D imaging in itself removes problems of surface discoloration. For the comparison of 3D images in general it may be fruitful to draw on technology from medical imaging.

Much material is known only from photographs, so that any successful technique will eventually have to be able to compare 2D and 3D images. There may be advantages here: for example, a 3D model might allow one to model lighting variability in 2D images. Perhaps the availability of some evidence in 3D will allow the 2D images to be reconstructed in 3D. Since the relationship of 2D to 3D images has been the subject of much exciting work described in this volume, it is apparent that there is scope for interesting work in this area.

From the numismatic point of view it is easy to imagine that different classes of coin may be better approached by different techniques. Thus it may be simpler to study large coins with relief on both sides in 2D, but smaller coins with incuse punchmarks in 3D. Whatever the case, it is certain that a portfolio of techniques will be required. A further practical concern is that, since coins often cannot be removed from museums, imaging equipment will need to be portable and relatively inexpensive. Technical problems will be demanding, but a start might be made by testing techniques on a small well-defined sequence of coins which display a range of image distortions.

One interesting consequence of such a project would be that automated die-studies could provide a control on those performed solely by human judgement. It is generally believed by numismatists that, although die studies (like most things) may be done badly, two adept practitioners will replicate each other's results. It would be interesting to test this by automated comparison, which will require the judgemental processes to be made explicit. It might also be helpful to museums to know whether 3D imaging has significant advantages over 2D imaging, before they invest in large-scale projects to image their collections. More than that, there is at least the potential for new imaging techniques to make possible die-studies on a scale not practical up to now, and thus to transform a subject.

Note

I am grateful to Alan Bowman, without whose prompting this paper would not have been written, and to Volker Heuchert, Henry Kim, and Greg Parker for useful discussion and practical help.

References

BEER TOBEY, L. and TOBEY, A. G. (1993), 'Experiments to simulate ancient Greek coins', in M. M. Archibald and M. R. Cowell, *Metallurgy in Numismatics*, vol. 3 (London): 28–35.

BLAND, R. F. (1991), 'The coinage of Gordian III from the mints of Antioch and Caesarea', unpublished doctoral thesis (Institute of Archaeology, University College, London).

CALLATAŸ, F. de. (1995), 'Calculating ancient coin production: seeking a balance', *Numismatic Chronicle* 155: 289–311.

CALLATAŸ, F. de. (1997), *Recueil quantitatif des émissions monétaires hellénistiques* (Wetteren).

CALLATAŸ, F. de., DEPEYROT, G. and VILLARONGA, L. (1993), *L'argent monnayé d'Alexandre le Grand à Auguste* (Bruxelles).

CRAWFORD, M. H. (1974), *Roman Republican Coinage* (Cambridge).

CRAWFORD, M. H. (1983), 'Numismatics', in M. H. Crawford (ed.), *Sources for Ancient History* (Cambridge): 185–233.

CRAWFORD, M. H. (1990), 'From Borghesi to Mommsen: the creation of an exact science', in M. H. Crawford, C. R. Ligota and J. B. Trapp (eds.), *Medals and Coins from Budé to Mommsen* (London): 125–32.

DOYEN, J.-M. (1989), 'L'atelier de Milan (258–268): recherches sur la chronologie et la politique monétaire des empereurs Valérien et Gallien', unpublished doctoral thesis (Université Catholique de Louvain).

DUNCAN-JONES, R. (1994), *Money and Government in the Roman Empire* (Cambridge).

GÖBL, R. (1978), *Antike Numismatik* (Munich).

HOWGEGO, C. (1990), 'Why did ancient states strike coins?', *Numismatic Chronicle* 150: 1–25.

HOWGEGO, C. (1992), 'The supply and use of money in the Roman world 200 BC to AD 300', *Journal of Roman Studies* 82: 1–31.

HOWGEGO, C. (1995), *Ancient History from Coins* (London).

KAENEL, H.-M. von (1991), ' "… ein wohl grossartiges, aber ausführbares Unternehmen": Theodor Mommsen, Friedrich Imhoof-Blumer und das Corpus Nummorum', *Klio* 73: 304–14.

KRAAY, C. M. (1956), *The Aes Coinage of Galba* (New York).

KRAAY, C. M. (1976), *Archaic and Classical Greek Coins* (London).

MØRKHOLM, O. (1982), 'A history of the study of Greek numismatics. III. c. 1870–1940: the scientific organisation', *Nordisk Numismatisk Årsskrift*: 7–26.

MØRKHOLM, O. (1991), *Early Hellenistic Coinage: From the Accession of Alexander to the Peace of Apamea (336–188 BC)* (Cambridge).

VOLK, T. R. (1987), 'Mint output and coin hoards', in G. Depeyrot, T. Hackens and G. Moucharte (eds.), *Rythmes de la production monétaire, de l'antiquité à nos jours* (Louvain-la-Neuve): 141–221.

WALKER, D. R. (1988), *Roman Coins from the Sacred Spring at Bath* (Oxford).

CHAPTER 12

Italian *Terra Sigillata* with Appliqué Decoration: Digitising, Visualising, and Web-Publishing

Eleni Schindler Kaudelka & Ulrike Fastner

Archaeologists today depend on the help of scientists for their work, but their relationship usually corresponds to the positions of consumer versus supplier, rather than the wish to form a working unit. The present paper is a short draft of a long-standing interdisciplinary project involving archaeology and photogrammetry, in which both sides tried to eliminate the gap between their disciplines. This co-operation led to a substantial extension of the timetable. Although the final report was completed according to the planned schedule after the immediate practical work on the project ceased and the administration was finished, neither party was happy with the results and could not consider the study complete. In fact, interdisciplinary work generally needs longer breaks between the different activity phases, and a considerable amount of supplementary meetings and conferences have to be planned. To emphasise this time-consuming aspect of the joint project, it is appropriate to record some steps of the project history.

Archaeological Context

Different types of red slip pottery occur all over the ancient Mediterranean world and are widely used for different purposes by archaeologists. Italian *terra sigillata* is considered as a very useful tool for dating evidence and as a basic indicator for trade patterns in early imperial times. If it comes in a mould-decorated technique, then it is a high-quality luxurious product treasured for its high level of aesthetic accomplishment. Plain vessels for everyday use are found even in remote provincial households. However, some of the plain tableware — cups, bowls, saucers, and also platters and plates — bear mould-made applied decorations on the rims.

The geographical boundaries of the present investigation were limited to the province of Noricum, and therefore all the results refer to the studied material found in Noricum up to 1999.

There is a 'bible' for Italian-style *sigillata*, the *Conspectus formarum terrae sigillatae italico modo confectae*,[1] in which C. Wells defines (p. 1) the material of the joint research programme as 'any Sigillata Loeschcke would have called "Arretine", any Sigillata, that is to say, wherever it is made, which in appearance, forms, colour, etc. closely resembles the products of the Arezzo workshops'.

There are some fifty types listed in the *Conspectus*, starting with late Republican rim forms in the tradition of black slip wares and covering the whole range of shapes until the end of the export of Italian *sigillata* in the mid-second century AD. Common shapes of *Applikensigillata* in Noricum are the bowls *Conspectus* 34, *acetabula* and *paropsides*, and the plates and platters *Conspectus* 20.4 (Figure 12.1). They occur all over the province, while the evidence for the different plates *Conspectus* 4, 6, and 21 is poor and mainly restricted to the Magdalensberg. No *Conspectus* 22 with appliqué decorations have so far been reported from Noricum.

Another quotation from the *Conspectus* deals with the applied decorations which are treated at the beginning of the study, of which P. M. Kenrick wrote (p. 149): 'The application of separately modelled motifs on the surface of a turned vessel by way of decoration goes back at least to the fourth century BC in black-glazed wares but the first use of the technique in Italian Sigillata production has all the appearance of a new invention.' So the newly invented decoration patterns consist mainly of rosettes, masks, *amorini*, and different mythological personages, dolphins, dogs, lions, and other animals, festoons, spirals, and vegetable and ornamental combinations of these, all in small dimensions and in a rather bad condition. *Applikensigillata* was mass-produced in the hasty, careless manner of assembly-line work. Most pieces are of poor quality and the longer the production lasts the worse the artefacts tend to be (Figure 12.2).

The fashion spread in Tiberian times from around AD 20 well into Trajanic times, around AD 115, and

Figure 12.1 Principal shapes of Italian *terra sigillata* with appliqué decoration occurring in Noricum.

---

[1] Ettlinger et al. (1990).

Figure 12.2 Poor quality of appliqué decorations due to careless handling.

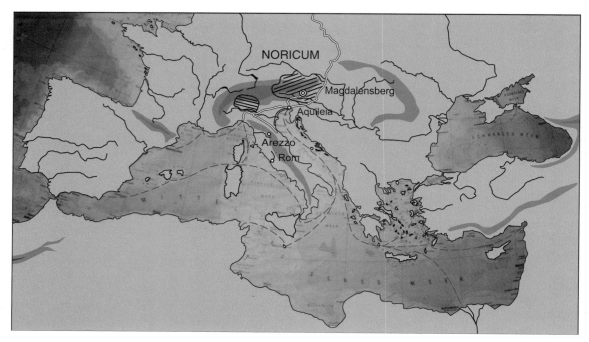

Figure 12.3 Map of the Mediterranean, drawn by K. Schindler; the hatched sections represent the production areas of Tardopadana *terra sigillata*.

lasted until the middle of the second century. Production, however, seemed to stop earlier, perhaps at the turn of the century. Appliqué decoration was first used in the workshops of Arezzo and later on adopted by several groups of central and south Italian manufacturers. It formed the main merchandise of the Tardopadana potters' group located in the north-western Po valley on the south fringe of the Alps. Distribution around the Mediterranean was quick and successful; *Applikensigillata* is well known in the chronologically relevant layers from Fréjus to Cherchel, from Corinth to Berenike, and from Aquileia to Monte Iato. Markets are divided, with the Tardopadana *terra sigillata* mainly being exported to the Danubian provinces, where it is considered the diagnostic find for Romanisation in newly conquered territories, while trade of the central and south Italian workshops was basically south- and east-bound (Figure 12.3).

The technique used provides a wide selection of decoration subjects, quasi-identical in shape and detail, but in different sizes. A master copy of each motif is used as a die to make a succession of impressions in a flat clay plaque. After firing, this *matrice a placca* serves as a mould

for the motifs which will be simply dried and applied to the vessels before coating and firing.[2]

A wide variety of motifs arranged in groups of six or eight, alternating in pairs on a rather small amount of different vessel shapes, restrict the decoration pattern, but nevertheless open up an interesting field of research. Regular symmetric compositions are normal, while the asymmetric ones are always accidental (Figure 12.4).

Archaeologists of the early twentieth century, mostly simply theoreticians, tended to establish wholly academic theories on the basis of one single piece at their disposal, while their familiarity with crafts remained poor. Practicability was almost never checked by experiment; ethno-archaeological approaches were unknown. These research methods led to a number of long-standing and therefore venerated theories, none of them closely scrutinised, and most of them simply incorrect.

Scholars were aware of the fact that copyright did not exist in antiquity, so that anybody could use a given product for his purposes. It was generally admitted that *surmoulage*, direct moulding from a vessel, was the

[2] Stenico (1954).

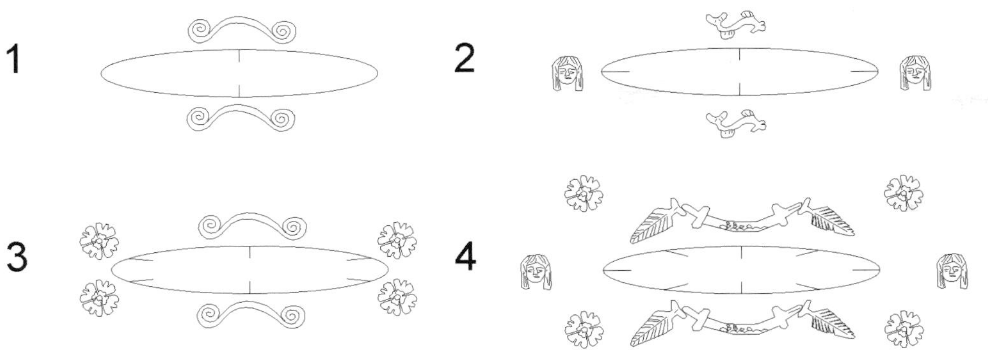

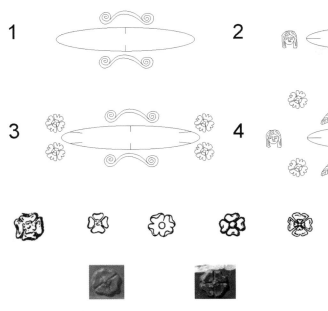

Figure 12.4 Composition patterns of Italian *terra sigillata* with appliqué decoration.

Figure 12.5 Different drawings of the same prototype in comparison with a photograph (Ohlenroth, 1934; Stenico, 1954; Vago, 1977; Gabler, 1973; Schindler, 1977).

preferred way of procuring new decoration types in a workshop. This would result in a slightly smaller die because of the shrinking of the clay and therefore diminishing sizes of appliqué decoration motifs would prove a process of deterioration over time, but also signify contacts between workshops, where the same models came into use. Nobody bothered about the surface coating that would ruin any detail of the shapes.

The National Science Foundation of Austria — Fonds zur Förderung der wissenschaftlichen Forschung in Österreich — granted support for a joint research programme in archaeology and photogrammetry in the years 1987–90 with Projects P6202 G and P87255 SPR. At the start of the investigation, the questions were simple:

1   Is it possible to raise the quality and/or the objectivity of drawings with technical measurement and drawing methods?
2   Is it possible to trace a generation line from the prototype of a die to the respective appliqué decoration on any vessel?
3   Is it possible to establish a basic pattern for a chronological approach?
4   Is it possible to define a typological set of dies for single potters?

The following was noted:

1   Archaeological results often depend on the comparability of drawings. A way to objectivate illustrations was badly needed (Figure 12.5).
2   The use of touch-free measurements to check the accurate sizes of the appliqué decorations was intended.
3   A series of chronologically ordered find-spots, the early Tiberian layers of the Magdalensberg, Flavia Solva (starting in early Flavian times), and those

layers of Carnuntum with a floruit in late Flavian times, was selected to examine the dating ranges (Figure 12.3).
4   For the creation of prototypes the sizes of the single appliqué decorations of all documented vessels of Noricum were to be compared (Figure 12.6).
5   All stamped plates, platters, and cups recorded from Noricum were included in the study.

The more common a type of pottery is, the less research interest it raises among ceramologists. The investigation of potsherds found day by day everywhere seems to lack the exotic. Lots of hypotheses have thus survived over almost a century and ultimately proved to be false. This is especially true for *Applikensigillata*. At the very first approach to the above-mentioned questions the 'state of the art' collapsed. Nothing seemed true any more. Nevertheless the project had to go on, primarily because the scientific part of the study produced important results right from the beginning.

Research Method

Photogrammetry is a method for the reconstruction of three-dimensional objects on the basis of photographs.[3] Aerial shots taken by special photogrammetric cameras are commonly used for standard applications. One of the most important and well-known applications consists in the production of topographic maps. The development of special close-range cameras permits measurements in different fields. Areas of application include architecture, building safety, the automotive industry, medicine, and so on. With the help of special algorithms and powerful software, standard 35 mm photographs can substitute for those produced by special cameras for geometrically correct reconstruction of three-dimensional objects. The importance of this method is increasing.

Photogrammetry can serve wherever touch-free, neutral, and independent measurements are required to accentuate objectivity and to restrict interpretation during documentation. Figure 12.7 illustrates the co-ordinate systems used in photogrammetry. The relation between image co-ordinate system and object co-ordinate system is given through the general equation

3   Kraus (1997).

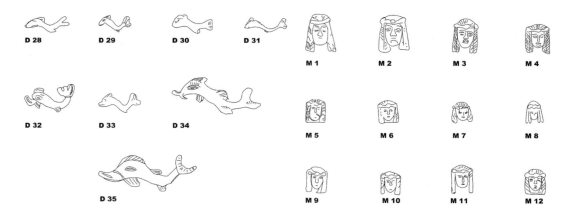

Figure 12.6 A series of dolphins and masks as examples for the creation of prototypes.

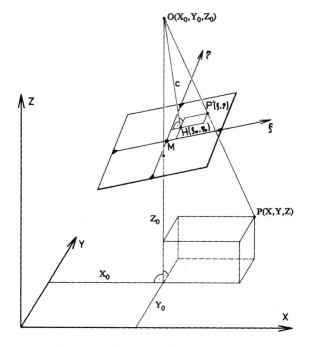

Figure 12.7 Principles of photogrammetry.

of projection:

$$\lambda^* \begin{pmatrix} \xi - \xi_0 \\ \eta - \eta_0 \\ -c \end{pmatrix} = R^{T*} \begin{pmatrix} X - X_0 \\ Y - Y_0 \\ Z - Z_0 \end{pmatrix}$$

where ξ, η are image co-ordinates; ξ_0, η_0, the principal point; λ, the stretching factor; R, the rotation matrix; X, Y, Z, the object co-ordinates; and X_0, Y_0, Z_0, the centre of projection. When the images are taken with a photogrammetric camera the interior orientation (principal point and focal length) is known exactly. A three-dimensional model is produced after relative orientation of the stereo pair. In order to calculate the co-ordinates in the object system, at least three control points are needed.

The number of unknown parameters will increase if a 35 mm camera is used instead of a photogrammetric device. The interior orientation is completely unknown, the camera is not very stable, and the geometry of the camera can change with each change of focus. The use of analytical or digital support allows the photogrammetric evaluation of amateur images, but more information 'from outside', that is to say a larger quantity of control points and special software for calculation of the camera parameters, will be necessary.

Requirements and Project Planning

The demands to be met were the following:

- close-range applications
- external control points
- high accuracy
- low cost.

The cheapest option available at the time for taking close-range photogrammetric images was to use a high-quality single lens reflex camera for 35 mm photographs. No special knowledge and training for comfortable handling and flexible use were necessary; therefore it provided a very efficient and fast way to take all the images. Nevertheless, special equipment and software for photogrammetric measurements were still required. M. Gruber made a first draft to check the accuracy of the photogrammetric measurement for small archaeological objects as early as 1985.

Setting up of the Experiment

From the start the following points emerged.

1 A large amount of very small pieces scattered in the storerooms of more than a dozen museums all over Noricum had to be evaluated. All pieces with appliqué decoration were arranged for photographs on a specially designed field of control points (Figure 12.8).[4] The images were taken as slightly convergent stereo pairs (angle about 15°) to ensure higher accuracy in measuring the depth. With this method the same accuracies in the x-, y-, and z-direction could be achieved.

---

[4] Gruber (1985).

Figure 12.8 Field of control points.

2 Easy transport of the camera and the field of control points was important. The three-dimensional field contains a sufficient number of control points for the determination of exterior orientation and information about camera parameters and stability of the camera. The co-ordinates of all its 74 control points could be measured with an accuracy of ± 20–$30\,\mu$m. Its size of about $20 \times 30 \times 10$ cm and the aluminium design, with the weight limited to 1.5 kg, are essential for use in a mobile laboratory.

3 The image scale had to be selected at a rate not smaller than 1 : 7 to yield correct reconversion to the natural size.

In this experiment the field of control points is not only used for camera calibration: it also offers the only possibility of bringing objective information into the fragments of *sigillata*.

Generally a field of control points is used for camera calibration. This means that focal length, principal point, and distortion of the lens will be calculated with the help of high-accuracy control points. At least 36 control points, symmetrical in the x-, y- and z-directions, are necessary to get good results in the calculation of the interior orientation.[5]

Photogrammetric Evaluation

The photogrammetric evaluation measurements were done with the Kern-DSR1 analytical plotter. The software, CRISP (close-range image set-up program), allowed the evaluation of non-metric images.[6] It was developed at the Institute for Image Processing and Digital Graphics of Joanneum Research, Graz, Austria, and specially adapted to meet the needs of small-dimension objects at the Institute for Applied Geodesy and Photogrammetry of the Technical University of Graz, Austria.

Provided an adequate number of control points exist, the calculation of the interior orientation of the camera

'on the job' is possible. Detailed investigations furnished evidence that the described method used for drawing the appliqué decoration has an accuracy of ± 0.05 mm, which greatly exceeds most of the conventionally made drawings or rubbings. There can be no doubt that photogrammetry is a very effective method of determining as exactly as possible the shape and size of appliqué decoration on Italian *sigillata*.

Data Storage

The photogrammetric measurements produced drawings to the scale of the original and files with all measured co-ordinates in a special data format. All images, negatives, plots, and co-ordinate files were put into archives together with all the archaeological data.

The archaeology/photogrammetry project seemed to be finished in 1989 but the handling of the data required familiarity with both the scientific and archaeological methods. The efficiency of the original method of data storage was felt to be provisional and did not satisfy all contributors.

For various reasons research was suspended for several years, and while everything was at a standstill, technical progress in data processing started to boom. It is interesting to recall that the first collection of data was kept on a series of 5¼ inch floppy disks, along with a container for photographic negatives and twelve card-index boxes. When work was resumed, it was decided to reprocess the old data. The new team had to cope with a completely changed electronic world but also with a new set of questions. Reorientation and the definition of a new objective were matters of necessity. At the start, a lot of supplementary work reinforced the decision to try to assemble all information in one framework which would be easy to handle.

The fact that more and more people are used to working with computer-aided-design systems on a personal computer and thus converting all drawings into a standard data format raised a new challenge: at this stage, with the important question about data formats, a large amount of time-consuming, enervating and boring transformations could not be avoided. The different data were stored in different formats, but it took the shrewdness of a detective to make the co-ordinates of a single piece available for consideration. Drawings of appliqué decorations existed as vector data, and the scanned vessel profiles were initially drawn using the classic lead wire technique and then digitised. These were stored as pixel data and the archaeological information was kept on Excel files, all of them in very long lists. They had to be combined into single files, each of them containing the profile of the fragment together with its appliqué decoration and linked to the archaeological catalogue and the photographs. The file formats used were:

1 GPI (general polygon interface): the only feasible medium for saving photogrammetric measurements, which contains headers with information about the

5 Böttinger (1981).
6 Fuchs and Leberl (1980).

points or lines and sets of three-dimensional co-ordinates. Easy to understand, but not read by commercial CAD software.

2 DXF (data exchange format): used as a standard in CAD, and therefore transformations had to be carried out.

3 TIFF (tagged image file format): one of the standards in digital image processing. All scanned profiles were stored as TIFF files and no problems arose in vectorising and converting them to DXF.

4 GIF (graphics interchange format): a bitmap format with high data compression; it is standard in all Internet browsers.

5 JPEG (Joint Photographic Experts Group): a standard image format with high data compression.

The steps shown in Figure 12.9 were completed to combine the information.

This was the first ceramological work in Austria not done with Chinese tracing ink, but precise computer-aided drawing of the profiles did not produce conclusive results. At present, too much time-consuming adjustment work is needed to make it efficient. The hand correction of the profile drawings and of some decorations was done in AutoCAD. The result produced high-quality life-size drawings, ready for printing. None of the original data proved to be unusable or outdated: competently gathered and documented data do not depend on a project or on the individual staff, for part of the staff had changed as well. Modern research can be done on the basis of old data. However, a considerable amount of energy and time (rather than money)

must be expended in order to read the data for further research.

Web-Publishing

When the web-publishing project was initiated, once more some unexpected work was added. Web browsers are standard on every computer, but CAD systems are less common. HTML (hypertext mark-up language) files were the best option for obtaining combinations of text, vector, and raster data on one common platform. Data formats with high compression rates which can be read on different systems were a matter of necessity. Today the different transformation steps that had to be undertaken may look rather complicated and unnecessary. But in the pioneer days of 1987, HTML and web-publishing were unknown. Whenever the project seemed finished, a new idea was born, which involved a lot of data manipulation to make further progress. Nevertheless, we are glad to be able to emphasise the fact that at any time all the old data could be used and no information which had been collected previously was lost.

The IAAD catalogue — *InterAktive Appliken-Datensammlung*, or InterActive Appliqué Data Collection (Figure 12.10) — contains a collection of photographs of all types of subjects in the dataset, and a map of Noricum showing all find-spots of Italian *terra sigillata* with appliqué decoration, allowing the user to search according to subject or location, respectively. Multiple links to each single piece with full information, such as inventory number, location of the piece, chronology, stamps,

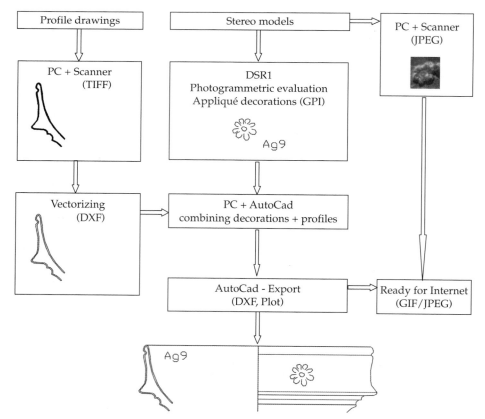

Figure 12.9 Diagram of data-transformation steps required to produce archaeological drawings ready for print.

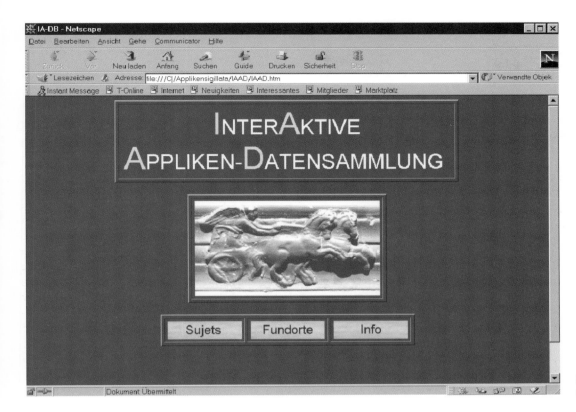

Figure 12.10 IAAD screenshot.

appliqué decoration, bibliographical references, and so on, offer a panorama of data which can serve as an index of Romanisation in Roman Noricum.

Prospects and Visions

More and better opportunities in hardware and software save a lot of time in data manipulation. A simple mouse-click is today's substitution for all our writing of transformation programs when exchanging standard data formats between different systems. New technical equipment is readily available but nevertheless there are still challenges to face and problems to solve while experimenting with new technologies.

An experimental set-up was arranged at the Technical University in Graz:

- a digital camera, Nikon Coolpix 950 (1600 × 1200 pixels)
- a field of high-accuracy control points[7]
- a personal computer with photogrammetric software, Photomod.

The use of a digital camera provides many advantages. Results are ready immediately and poor-quality images can be redone at once—a rather big advantage if one needs to travel from one museum to another to gather the necessary data. Delays due to the processing of the prints in the photographic laboratory no longer have to be taken into account when planning the timetable, since modern software can import digital images directly. No scanning will be necessary and therefore no danger of any loss of quality exists.

Unfortunately there is a downside, as ever. The geometry of digital cameras is not very stable, automatic focusing poses a problem for camera calibration, and tests showed a large degree of image distortion, especially on the outer edges of the picture.

The test involved placing all pieces with appliqué decoration in a single layer as close as possible to the central focus of the field of control points to eliminate as far as possible the problem of distortion.

The example shows more than 50 control points—enough points to calculate orientation of the stereo pair and to determine the unknown parameters of the camera. All measurements were done on a personal computer using the commercial photogrammetric software Photomod. This software allows us to use stereo pairs without any knowledge of the camera parameters, provided that the number of control points is adequately high. Automatic correlation of the image orientation is a big advantage. Furthermore, a semi-automatic correlation mode provides assistance for the exact positioning of the measuring mark for the stereoscopic measurements. In this case plotter evaluation is no longer necessary, since the calculations are done on the personal computer. It is necessary to use anaglyphic glasses while doing stereo drawing, and this results in a less comfortable working position than when doing the calculation work on an analytical plotter, for example, the Kern-DSR1.

The accuracy test ended satisfactorily. Only very slight differences can be observed in the overlays in comparison with the original drawings. Repetition of the experiment under the same conditions with different persons doing the evaluations and the measurements

[7] Schwendtner (1995).

0 5 mm

Figure 12.12 Manufacturing differences of appliqué decorations.

Figure 12.11 Accuracy test comparing a stereo image pair photographed with a 35 mm camera and with a digital camera.

showed all lines closely corresponding and accuracy clearly exceeding that of a parallel experiment done with handmade drawings (Figure 12.11).

Archaeological Evaluation

The *Applikensigillata* project produced five main archaeological results.

The first target was to get correct drawings by photogrammetric evaluation in order to compare different sizes of similar decorations. It can be assumed that the same motif applied four times on one platter comes from one *matrice a placca* designed with a single die. The differences in the drawings are technological accidents and not intentional stylistic variations. They result from careless handling during the different production steps (Figure 12.12). Therefore only one drawing will figure on the prototype plates.

An unexpected reorientation proved to be extremely favourable. While the data over the museums of Roman Austria were being compiled, staff there had asked for a comprehensive catalogue of all the finds of Italian *sigillata* with appliqué decorations in Noricum. Although no general catalogue was planned at the beginning of the study, the IAAD does now offer this, besides being an open database suitable for updating. A preliminary repertoire of the decoration of the potter L. Gellius was established as a first example, but the evidence of the finds in Noricum was not sufficient to define it, and more information from a 1982 study covering the whole empire had to be added.[8]

A series of chemical analyses and thin sections cross-read with the stamps opened up a new field of results reflecting quality and repertoire both in shapes and decorations, but mainly in provenance questions.[9] It was very helpful for the development of typology, while at the same time new refinement of the chronological issues could be achieved.

Good results on the distribution patterns of *terra sigillata* with appliqué decoration in Noricum, and therefore data for the study of the Romanisation, have been obtained, while once again the old myth of a cartel-like organisation of Roman craft and trade was proved false.

Last but not least, we took care to eliminate any surviving old catalogue based on the observation of a single piece. One of the referees for the manuscript

characterised it as the most crucial negative result of the last ten years. But it takes some courage to publish negative results of research without offering new hypotheses. *Italische Terra Sigillata mit Appliken in Noricum* was published in 2001 by the Austrian Academy of Science (Österreichische Akademie der Wissenschaften), with a CD-ROM containing the IAAD catalogue.

It is a pleasure to communicate that the story is not over. A young scientist from Rheinzabern is about to start a big research programme on the Ludovici collection in the Speyer Museum. The method of digitising and visualising described here seems to represent the appropriate method for the questions that arise when comparing dies, stamps, and mould-made decorations. Today's possibilities will replace outdated devices, so a digital camera instead of a 35 mm single reflex lens, and a laptop instead of a card index box, will accompany the dataset of control points along with the basics of interdisciplinary experience.

This unexpectedly episodic novel proved to be most rewarding and gratifying: even with the overwhelming breakneck speed in technological advance, careful and thorough reflection aimed at finding adequate investigation methods did not become outdated within fourteen years, while modifications that are simple and easy to carry out indicate research that will subsequently bear fruit. And it is comforting to be reminded that archaeological work cannot always be assessed by the same criteria used in evaluating the efficiency of technical science.

References

BÖTTINGER, W. U. (1981), *Theoretische und experimentelle Untersuchungen zur Genauigkeit der Nahbereichsphotogrammetrie* (München).

ETTLINGER, E., et al. (1990), *Conspectus formarum terrae sigillatae italico modo confectae* (Bonn).

FUCHS, H. and LEBERL, F. (1980), *'CRISP', a Software Package for Close Range Photogrammetry for the Kern DSR1* (Aarau).

GABLER, D. (1973), *Italische Sigillaten in Nordwestpannonien*, Wissenschaftliche Arbeiten aus dem Burgenland 51 (Eisenstadt).

GRUBER, M. (1985), *Nahbereichsphotogrammetrie mit 35-mm Film und archäologischen Objekten*: *Seminararbeit an der TU Graz*.

GRUBER, M. (1991), 'Zur photogrammetrischen Dokumentation von archäologischen Kleinfunden', *XXVIII Commission V Congress of ISPRS Zürich 1990*: 234–7.

GRUBER, M. and SCHINDLER KAUDELKA, E. (1990), 'Photogrammetrische Dokumentation von italischer

---

[8] Zabehlicky-Scheffenegger (1982).
[9] Schindler Kaudelka et al. (1998).

Terra Sigillata mit Appliken', *Mitteilungen der geodätischen Institute der TU Graz* 69: 129–33.

KRAUS, K. (1997), 'Photogrammetrie', Band 1, *Grundlagen und Standardverfahren* 6 (Bonn).

OHLENROTH, L. (1934), 'Italische Sigillata mit Auflagen aus Rätien und dem römischen Germanien', *24–25. Berichte der römisch-germanischen Kommission*: 234–54.

SCHINDLER, M. and SCHEFFENEGGER, S. (1977), *Die glatte rote Terra Sigillata vom Magdalensberg* (Klagenfurt).

SCHINDLER KAUDELKA, E., FASTNER, U. and GRUBER, M. (1998), 'Note sur les sigillées italiques à décor appliqué', *SFECAG, Actes du Congrès d'Istres*: 253–64.

SCHINDLER KAUDELKA, E., FASTNER, U. and GRUBER, M. (2001), *Italische Terra Sigillata mit Appliken in Noricum* (Wien).

SCHINDLER KAUDELKA, E., SCHNEIDER, G. and ZABEHLICKY-SCHEFFENEGGER, S. (1997), 'Les sigillées padanes et tardo-padanes: nouvelles recherches en laboratoire', *SFECAG, Actes du Congrès du Mans*: 481–94.

SCHWENDTNER, K. (1995), 'Testfeldkalibrierung für Teilmess- und CCD-Kameras', dissertation (University of Graz).

STENICO, A. (1954), 'Matrici a placca per applicazioni di vasi aretini del Museo Civico di Arezzo', *Archeologia Classica* 6: 43–76.

VAGO, E. B. (1977), 'Die oberitalisch-padanische Auflagen-Sigillata in Transdanubien', *Acta Academiae Scientiarum Hungaricae* 29: 77–124.

ZABEHLICKY-SCHEFFENEGGER, S. (1982), 'Die Geschäfte des Herren Lucius G.', *RCRF Acta* 21–2: 105–16.

CHAPTER 13

Shape from Profiles

Roberto Cipolla & Kwan-Yee K. Wong

I. Introduction

Profiles (also known as outlines, apparent contours or silhouettes) are often a dominant feature in images. They can be extracted relatively easily and reliably from the images, and provide rich information about both the shape and motion of an object. Classical techniques[1] for model reconstruction and motion estimation depend on point or line correspondences, and hence cannot be applied directly to profiles, which are viewpoint dependent. This calls for the development of a completely different set of algorithms specific to profiles. This paper will give a brief review of some state-of-the-art algorithms for model building and motion estimation from profiles alone.

II. Profiles of Surfaces

Profiles are projections of contour generators which divide the visible from the occluded part of the surface[2] and hence depend on both the surface geometry and camera position. In general, two contour generators on a surface, associated with two different camera positions, will be two distinct space curves, and thus the corresponding profiles on the images do not readily provide point correspondences which are needed to estimate the camera motion. A *frontier point*[3] is the intersection of two contour generators and lies on an epipolar plane tangent

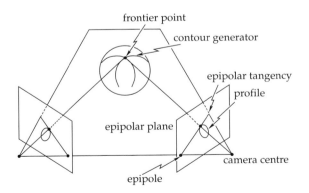

Figure 13.1 A frontier point is the intersection of two contour generators and lies on an epipolar plane which is tangent to the surface. It follows that a frontier point will project to a point on the profile which is also on an epipolar tangent.

[1] Faugeras (1993).
[2] Cipolla and Giblin (1999).
[3] Ibid.

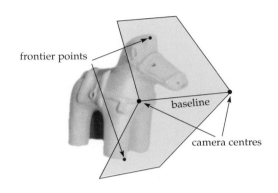

Figure 13.2 The outer epipolar tangents correspond to the two epipolar tangent planes which touch the object, and are always available in any pair of views except when the baseline passes through the object.

to the surface (see Figure 13.1). It follows that a frontier point will project to a point on the profile which is also on an epipolar tangent.[4] Epipolar tangencies thus provide point correspondences on profiles, and can be exploited for motion estimation (see Figures 13.2 and 13.3).[5]

III. Motion Estimation

A practical algorithm for motion estimation from profiles, in the case of complete circular motion (where an object is rotated about a fixed axis in front of an uncalibrated camera), was introduced in Mendonça et al. (2001). In Wong and Cipolla (2001) the profiles from (incomplete) circular motion were exploited for the registration of any arbitrary general view, and a complete system for model acquisition from uncalibrated profiles under both circular and general motion was developed. A summary of these techniques[6] is given below.

The three main image features in circular motion (Figure 13.4), namely the image of the rotation axis $\mathbf{l}_s$, the horizon $\mathbf{l}_h$, and a special vanishing point $\mathbf{v}_x$,[7] are fixed throughout the sequence and satisfy

$$\mathbf{v}_x \cdot \mathbf{l}_h = 0 \quad \text{and} \tag{13.1}$$

$$\mathbf{v}_x = \mathbf{KK}^\mathrm{T} \mathbf{l}_s \tag{13.2}$$

where $\mathbf{K}$ is the 3×3 camera calibration matrix. The camera motion (which can be conveniently represented by

[4] Porrill and Pollard (1991).
[5] See Cipolla and Giblin (1999) for a review of previous work.
[6] Mendonça et al. (2001).
[7] See Mendonça et al. (2001) for details.

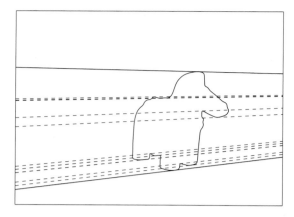
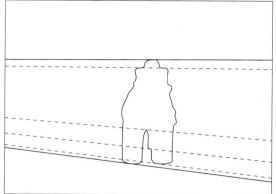

Figure 13.3 Two discrete views showing seventeen epipolar tangents in total, of which only four pairs are in correspondence. The use of the two outer epipolar tangents (in solid lines), which are guaranteed to be in correspondence, avoids false matches due to self-occlusions, and greatly simplifies the matching problem.

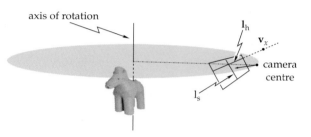

Figure 13.4 Under circular motion the image of the rotation axis l_s, the horizon l_h, and a special vanishing point v_x are fixed throughout the sequence.

the epipolar geometry and the fundamental matrix) can be parametrised explicitly in terms of these features,[8] and the fundamental matrix is given by

$$\mathbf{F} = [\mathbf{v}_x]_\times + \kappa \tan \frac{\theta}{2}(\mathbf{l}_s \mathbf{l}_h^T + \mathbf{l}_h \mathbf{l}_s^T) \qquad (13.3)$$

where θ is the angle of rotation, and κ is a constant which can be determined from the camera intrinsic parameters. A sequence of N images taken under circular motion, with known camera intrinsic parameters, can hence be described by $N + 2$ motion parameters. By using the two outer epipolar tangents,[9] the N images will provide $2N$ (or 2 when $N = 2$) independent constraints on these parameters, and a solution will be possible when $N \geq 3$.

The circular motion will generate a *web of contour generators* around the object (see Figure 13.5), which can be exploited to register any new arbitrary general view. Given an arbitrary general view, the associated contour generator will intersect with this web and form frontier points. If the camera intrinsic parameters are known, the six motion parameters of the new view can be fixed if there are six or more frontier points on the associated contour generator. This corresponds to having a minimum of three views under circular motion, each providing two outer epipolar tangents to the profile in the new general view (see Figure 13.6).

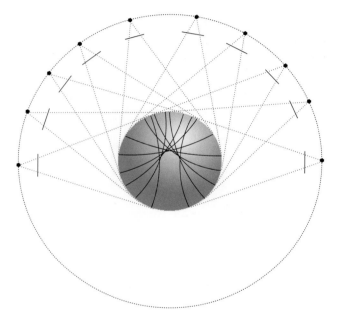

Figure 13.5 The circular motion will generate a web of contour generators around the object, which can be used to register any new arbitrary general view.

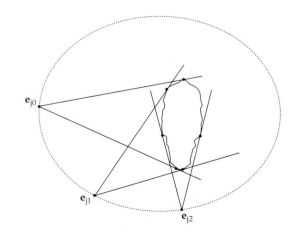

Figure 13.6 Three views from circular motion provide six outer epipolar tangents to the profile in the new general view for estimating its pose.

---

[8] Viéville and Lingrand (1996).
[9] Wong and Cipolla (2001).

The motion estimation proceeds as an optimisation which minimises the reprojection errors of epipolar tangents. For view i and view j, a fundamental matrix $\mathbf{F}_{ij}$ is formed from the current estimate of the motion parameters, and the epipoles $\mathbf{e}_{ij}$ and $\mathbf{e}_{ji}$ are obtained from the right and left nullspaces of $\mathbf{F}_{ij}$. The outer epipolar tangent points $\mathbf{t}_{ij0}$, $\mathbf{t}_{ij1}$ and $\mathbf{t}_{ji0}$, $\mathbf{t}_{ji1}$ are located in view i and j respectively (see Figure 13.7). The reprojection errors are then given by the geometric distances between the epipolar tangent points and their epipolar lines

$$d_{ijk} = \frac{\mathbf{t}_{jik}^{\mathrm{T}}\mathbf{F}_{ij}\mathbf{t}_{ijk}}{\sqrt{(\mathbf{F}_{ij}\mathbf{t}_{ijk})_1^2 + (\mathbf{F}_{ij}\mathbf{t}_{ijk})_2^2}} \tag{13.4}$$

$$d_{jik} = \frac{\mathbf{t}_{jik}^{\mathrm{T}}\mathbf{F}_{ij}\mathbf{t}_{ijk}}{\sqrt{(\mathbf{F}_{ij}^{\mathrm{T}}\mathbf{t}_{jik})_1^2 + (\mathbf{F}_{ij}^{\mathrm{T}}\mathbf{t}_{jik})_2^2}}. \tag{13.5}$$

For a sequence of N images taken under circular motion, the image of the rotation axis and the horizon are initialised approximately, and the angles are arbitrarily

initialised. The cost function is given by

$$C_{cm}(\mathbf{x}) = \frac{1}{4}\sum_{i=1}^{N}\sum_{j=i+1}^{min(j+3,N)}\sum_{k=1}^{2} d_{ijk}(\mathbf{x})^2 + d_{jik}(\mathbf{x})^2 \tag{13.6}$$

where $\mathbf{x}$ consists of the $N + 2$ motion parameters.

For arbitrary general motion, the six motion parameters can be initialised by roughly aligning the projection of the 3D model built from the estimated circular motion with the image. The cost function of general motion for view j is given by

$$C_j(\mathbf{x}') = \sqrt{\frac{\sum_{i=1}^{N} f_{ij}\sum_{k=1}^{2} d_{ijk}(\mathbf{x}')^2 + d_{jik}(\mathbf{x}')^2}{4\sum_{i=1}^{N} f_{ij}}} \tag{13.7}$$

where $\mathbf{x}'$ consists of the six motion parameters; f_{ij} is 0 if the baseline formed with view i passes through the object, otherwise it is 1.

The motion estimation procedure is summarised in algorithm 1.

IV. Model Reconstruction

The image profiles of an object provide rich information about its shape. Under known camera motion, it is possible to reconstruct a model of the object from its profiles. For continuous camera motion and simple smooth surfaces, a surface representation can be obtained from the profiles using the epipolar parametrisation.[11] Cipolla and Blake developed a simple numerical method for estimating depth from a minimum of three discrete views by determining the osculating circle on each epipolar plane. Vaillant and Faugeras (1992) developed a similar algorithm which uses the radial plane instead of the epipolar plane. Boyer and Berger (1997) derived a depth formulation from a local approximation of the

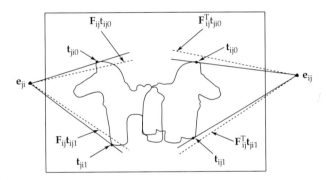

Figure 13.7 The motion parameters can be estimated by minimising the reprojection errors of epipolar tangents, which are given by the geometric distances between the epipolar tangent points and their epipolar lines.

Algorithm 1 Motion estimation from profiles

Extract the profiles using cubic B-spline snakes[10]
initialize $\mathbf{l}_s$, $\mathbf{l}_h$ and the $N - 1$ angles for the circular motion
while not converged **do**
    compute the cost for circular motion using (13.6)
    update the $N + 2$ motion parameters to minimize the cost
end while
Form the essential matrices from the fundamental matrices using the calibration matrix
Decompose the essential matrices to obtain the projection matrices
Build a partial model using an octree carving algorithm
for each arbitrary general view **do**
    initialize the six motion parameters using the partial model from circular motion
    while not converged **do**
        compute the cost for the general motion using (13.7)
        update the six motion parameters to minimize the cost
    end while
end for
Refine the model using the now calibrated images from general motion.

---

[10] Cipolla and Blake (1992).
[11] Cipolla and Blake (1992).

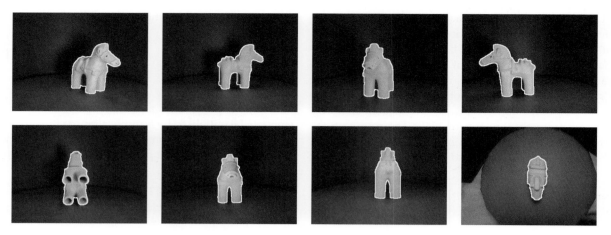

Figure 13.8 Top: four images from an uncalibrated sequence of a haniwa taken under circular motion (views 1, 4, 7, and 10, respectively). Bottom: four images from arbitrary camera positions (views 12, 13, 14, and 15, respectively).

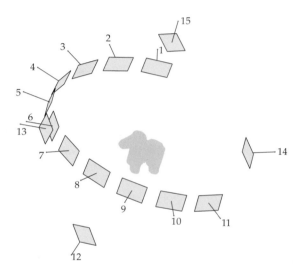

Figure 13.9 Camera poses estimated from the sequence of the haniwa.

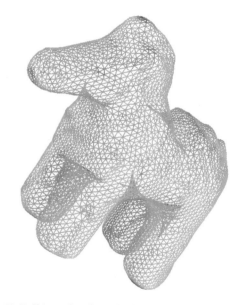

Figure 13.10 Triangulated mesh of the haniwa model, composed of 12,028 triangles.

surface up to order 2, which allows the local shape to be estimated from three consecutive views by solving a pair of simultaneous equations.

Alternatively, for discrete motion and objects with more complex geometry, a volumetric model can be obtained by an octree carving algorithm.[12] This technique was chosen in this section to illustrate the results of reconstruction using the motion estimated from profiles, and thus is described in more detail here. Initially, the octree consists of a single cube in space which encloses the model to be reconstructed. The cube is projected onto each image and classified as either (a) completely outside one or more profiles, (b) completely inside all the profiles, or (c) ambiguous. If the cube is classified as type (c), it is subdivided into eight sub-cubes (hence the name octree), each of which is again projected onto the images and classified. This process is repeated until a preset maximum level (resolution) is reached. Cubes classified as type (a) are thrown away, leaving type (b) and (c) cubes, which constitute the volumetric model of the object. Surface

triangles, if needed, can be extracted from type (c) cubes using a marching cubes algorithm.[13] The octree carving technique is summarised in algorithm 2, and the implemented algorithm for octree carving can be downloaded from: http://svr-www.eng.cam.ac.uk/~cipolla.

It is worth noting that both the surface and volumetric models, estimated only from the profiles of the object, correspond to the visual hull of the object with respect to the set of camera positions. Concavities cannot be recovered as they never appear as part of the profiles. In order to 'carve' away the concavities, methods such as space carving[14] could be used instead.

Figure 13.8 shows eight images of a sequence of fifteen used to recover the geometry of a Japanese archaeological statue (haniwa). The profiles are outlined and only these are used to estimate the camera motion. The recovered camera motion using only the profiles is shown in Figure 13.9, while Figure 13.10 shows the wire

---

[12] Szeliski (1993).

[13] Lorensen and Cline (1987).
[14] Kutulakos and Seitz (2000); Broadhurst et al. (2001).

Algorithm 2 Octree carving from profiles

Initialize a cube that encloses the model
while max level not reached **do**
    for each cube in the current level **do**
        project the cube onto each image
        classify the cube as either:
            (a) completely outside one or more profiles,
            (b) completely inside all the profiles, or
            (c) ambiguous
        if the cube is classified as type (c) **then**
            subdivide the cube into eight sub-cubes
            add the sub-cubes to the next level
        end if
    end for
    Increase the level count
end while

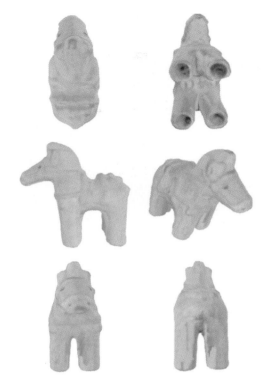

mesh reconstructed from the motion estimated using profiles. Figure 13.11 shows views of the model.

In another experimental sequence, fourteen images of an outdoor sculpture were acquired by a hand-held camera (see Figure 13.12). An approximate circular motion of the camera was achieved by using a string which was fixed to the ground by a peg at one end. A circular path on the ground was then obtained by rotating the free end of the string about its fixed end. Each image in the sequence was acquired by positioning the camera roughly at a fixed height above the free end of the (rotating) string, and pointing it towards the

Figure 13.11 Refined model of the haniwa after incorporating the four arbitrary general views. The model is now fully covered with textures and shows great improvements in shape, especially in the front, back, and top views.

Figure 13.12 Eight images of an outdoor sculpture acquired by a hand-held camera. Although the camera centre roughly followed a circular path, the orientation of the camera was unconstrained and hence the image of the rotation axis and the horizon were not fixed throughout the image sequence.

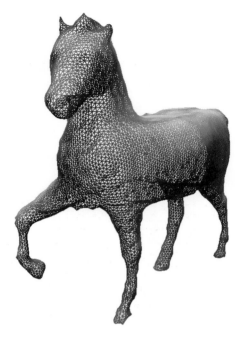

Figure 13.13 Triangulated mesh of the outdoor sculpture of a horse. This model was composed of 29,672 triangles.

sculpture. Note that since the camera centre, the string, and the rotation axis were roughly coplanar, the image of the string in each image provided a very good estimate for the image of the rotation axis $\mathbf{l}_s$. Although the camera centre roughly followed a circular path, the orientation of the camera was unconstrained, and hence the image of the rotation axis $\mathbf{l}_s$ and the horizon $\mathbf{l}_h$ were not fixed throughout the image sequence. In order to allow the camera motion to be estimated using the circular motion algorithm, the images were first rectified so that the image of the string (i.e. the image of the rotation axis) became a fixed vertical line passing through the principal point throughout the sequence. A transformation induced by a rotation about the x-axis of the camera was then applied to each image so that the image of the fixed end of the string became a fixed point on $\mathbf{l}_s$ throughout the rectified sequence. The resulting image sequence resembled a circular motion sequence, in which the horizon $\mathbf{l}_h$, the image of the rotation axis $\mathbf{l}_s$, and the special vanishing point $\mathbf{v}_x$ were fixed. The algorithm for circular motion estimation was then applied to this rectified sequence, and the resulting camera poses were then iteratively refined by applying the general motion algorithm. The 3D model built from the estimated motion is shown in Figure 13.13.

V. Conclusions

The incorporation of arbitrary general views reveals information which is concealed under circular motion, and greatly improves both the shape and texture of the 3D models. Since only profiles have been used in both

the motion estimation and octree carving, no corner detection or matching is necessary. This means that the algorithm is capable of reconstructing any kind of objects, including *smooth* and *textureless* surfaces. Experiments on various objects have produced convincing 3D models, demonstrating the practicality of the algorithm.

References

BOYER, E. and BERGER, M. O. (1997), '3D surface reconstruction using occluding contours', *International Journal of Computer Vision*, 22: 219–33.

BROADHURST, A., DRUMMOND, T. and CIPOLLA, R. (2001), 'A probabilistic framework for space carving', *Eighth IEEE International Conference on Computer Vision, (ICCV-01), July 7–14, 2001, Vancouver, British Columbia, Canada*, IEEE Computer Society, vol. 1: 388–93.

CIPOLLA, R. and BLAKE, A. (1992), 'Surface shape from the deformation of apparent contours', *International Journal of Computer Vision*, 9: 83–112.

CIPOLLA, R. and GIBLIN, P. J. (1999), *Visual Motion of Curves and Surfaces* (Cambridge, UK).

FAUGERAS, O. D. (1993), *Three-Dimensional Computer Vision: a Geometric Viewpoint* (Cambridge, MA, and London).

KUTULAKOS, K. N. and SEITZ, S. M. (2000), 'A theory of shape by space carving', *International Journal of Computer Vision*, 38: 197–216.

LORENSEN, W. E. and CLINE, H. E. (1987), 'Marching cubes: a high resolution 3D surface construction algorithm', in M. C. Stone (ed.), *SIGGRAPH 87: Computer Graphics Proceedings* (Anaheim, CA): 163–70.

MENDONÇA, P. R. S., WONG, K.-Y. K. and CIPOLLA, R. (2001), 'Epipolar geometry from profiles under circular motion', *IEEE Transactions on Pattern Analysis and Machine Intelligence*, 23/6: 604–16.

PORRILL, J. and POLLARD, S. B. (1991), 'Curve matching and stereo calibration', *Image and Vision Computing*, 9: 45–50.

SZELISKI, R. (1993), 'Rapid octree construction from image sequences', *CVGIP: Image Understanding*, 58: 23–32.

VAILLANT, R. and FAUGERAS, O. D. (1992), 'Using extremal boundaries for 3D object modelling', *IEEE Transactions on Pattern Analysis and Machine Intelligence*, 14: 157–73.

VIÉVILLE, T. and LINGRAND, D. (1996), 'Using singular displacements for uncalibrated monocular visual systems', in B. Buxton and R. Cipolla (eds.), *Computer Vision: ECCV '96. 4th European Conference on Computer Vision, Cambridge, UK*, vol. 2: 207–16.

WONG, K.-Y. K. and CIPOLLA, R. (2001), 'Structure and motion from silhouettes', in *Eighth IEEE International Conference on Computer Vision (ICCV-01), July 7–14, 2001, Vancouver, British Columbia, Canada*, IEEE Computer Society.

CHAPTER 14

The Skull as the Armature of the Face: Reconstructing Ancient Faces

R. A. H. Neave & A. J. N. W. Prag

A surgeon operating on the face removes layers of skin and tissue to reveal the structure of the muscles and then the hard, bony skull beneath. The medical artist does the reverse: starting from the skull he builds up each muscle in turn using well-established statistics for flesh thickness, and then adds a layer of clay for the subcutaneous tissue and skin. If one thinks of the skull as resembling the steel armature of a modern building, then the muscles and their attachments represent the service ducts and the skin the final cladding of concrete or brick. It is the proportions of the underlying armature — the skull — that dictate the form of the building, or the face, and give it its individuality.

This sounds both simple and obvious, but there are of course some caveats. We have described the technique of facial reconstruction on several occasions, perhaps most accessibly in chapter 2 of *Making Faces*, a book which is fundamental to this whole study.[1] One never works from the actual skull, in either forensic or archaeological cases, because it is unique and must not be exposed to the risk of damage. The first step is to make a plaster cast that is an exact replica of the original. In some instances the skull is not immediately accessible, for example where it is still covered by tissue or burial wrappings, as on the bog bodies of northern Europe or with Egyptian mummies: here the practice (described in greater detail by Linney et al. in Chapter 15) is now to make a series of CT scans from which an image of the skull can be recreated by a computer program; from these a further program will make a plastic replica either by carving the shape out from a solid block, or by using the technique of selective laser-sintering where a computer-guided laser beam is used to polymerise a powdered plastic. This replica then serves as the basis for the facial reconstruction in the same way as the plaster cast.

Pegs are inserted into the cast at a given number of points, varying according to individual practice from twenty-one to thirty-four, to mark the thickness of the flesh, using statistics that have been built up for over a century. At first they were based on physical measurements taken from cadavers — a pin pushed through a rubber marker was inserted into the face and the marker adjusted to mark the depth to which the pin had penetrated. For the last twenty-odd years the measurements have been taken from living people using ultrasonic techniques. Sex, race, and age are all relevant, and must

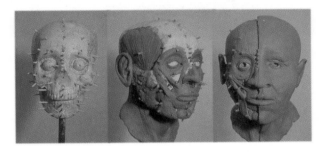

Figure 14.1 Three stages in the reconstruction of the face of the Copenhagen mummy. (a) Cast of skull with pegs and replacement eyeballs inserted. (b) The major muscles, mouth, nose, and ears in place. (c) Half the head still bare of the final covering of subcutaneous tissue and skin, revealing the structure of the muscles underneath.

be taken into account. The exact method by which the 'flesh' is applied varies in different countries. That found most reliable in Britain and in Russia (where it was pioneered in its modern form by Mikhail Gerasimov) is to reconstruct the main muscles or muscle groups of the face in clay.[2] It is important to take into account the positions and strength of the main muscle-insertions, for these give an indication of the strength of the muscles and thus of their bulk. The exact size of the muscles is less important, for the final fatness or thinness of the face is determined by the thickness of the overlying soft tissue (represented on the reconstruction by a final layer of clay), but like the service ducts on a building it is the position, the direction of pull, and the approximate strength of the muscles that mould its shape (Figure 14.1). The alternative method, favoured in the USA, is simply to cover the skull with pellets of clay up to the level of the pegs without regard to the subtle modelling dictated by the muscular structure, resulting in a much more wooden and lifeless end-product.

Thus far the whole process is dictated by the form of the skull and the medical and pathological evidence provided by the skull and the postcranial skeleton. In forensic cases and in the case of bog bodies and mummies there is sometimes further evidence about the dead person's hair (and where appropriate his beard) and perhaps some other details such as spectacles or tattoos. However, if only the skeleton remains one can only turn to external evidence, such as portraits, for this information

[1] Prag and Neave (1997), 20–40.

[2] Gerasimov (1971).

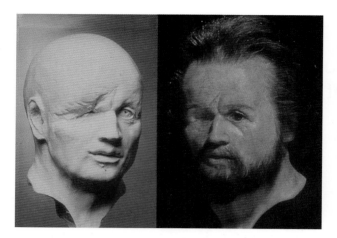

Figure 14.2 The reconstructed head of Philip II of Macedon (382–336 BC): (a) without and (b) with hair and beard.

Figure 14.3 The reconstructed head of Karen Price, and a photograph taken when she was alive.

once an identification has been made. Thus in the case of Philip II of Macedon what convinced the excavator was a photograph of a hairless and colourless grey head made of clay which demonstrated the blind right eye that was the proof of his identity (Figure 14.2(a)). What the public were shown later was a waxwork which depicted him with hair and a short beard based on the few surviving portraits of Philip, for by then the identity of the skull and the tomb had been proved — more on this below (Figure 14.2(b)). It is very important to understand, as some critics do not, that a facial reconstruction which was used to prove identity purely on the grounds that it resembled a portrait would be open to a charge of circularity, quite apart from the fact that it would be impossible to show that the medical artist had not been influenced by the portrait. Although facial reconstruction is now acceptable evidence of identification in a court of law, it must be supported by at least one further piece of evidence before it becomes proof. The remains of Karen Price, found in a garden in Cardiff in 1989, were first identified when a reconstruction of her face was shown on *Crimewatch UK*, which could be compared with photographs of the girl in life, but it needed comparison of her dental records with the dentition of the skull to provide the confirmatory proof (Figure 14.3).

The work is painstaking, and requires a knowledge of pathology, anatomy, dentistry, and much more to build up a full case history — indeed, it cannot be done properly except as a piece of teamwork (the parallel with the building, its architect, and his or her team of contractors and subcontractors can be continued). Therein lies its fascination for both practitioner and viewer, and therein lies its scientific value. The resulting face will not be a portrait *sensu stricto*, for that involves a knowledge of the person's character, betrayed by small details such as laughter lines that leave no evidence on the bone; but it will be a recognisable likeness, in effect a three-dimensional report bringing together all the available evidence garnered by the research team. One might argue that the extra time, effort, and expertise required to produce this three-dimensional report were unnecessary

and that a conventional written report illustrated with photographs and drawings would serve the purpose equally well. In some cases this is no doubt true, particularly since over the last twenty years both scholars and the general public have become accustomed to seeing facial reconstructions on television, in books and articles, and in museums. None the less, a bald black-and-white report can never carry the impact of the face itself. This was illustrated for us most impressively by the first presentation of the reconstruction of the face of Philip II of Macedon. Two colleagues, a plastic surgeon and a facio-maxillary surgeon, recognised the blinding injury to the right eye and thus enabled us to establish the identity of the person buried in tomb II at Vergina as King Philip II, who was known from the historical sources to have suffered just such an injury during the siege of Methone in 354 BC, eighteen years before his death. Naturally we passed the information immediately to the excavator, Professor Manolis Andronicos, in writing. In writing back to us, he sounded reasonably convinced. Yet it was only when we presented him with the photographs of the reconstruction which clearly showed the eye-wound that he allowed himself and his excavation team to believe it, and finally committed himself publicly to the identification for which he had always hoped. We ourselves showed the pictures two weeks later in a lecture to the 12th International Congress of Classical Archaeology in 1983, held appropriately enough in Athens: neither of us will forget the simultaneous intake of breath by several hundred of our professional colleagues when the disfigured face appeared on the screen.

A second more technical but no less important point lies in the fact that making the reconstruction in this way is a three-dimensional process: not only must the face fit, but the parts of the skull must fit. Staying with the case of Philip of Macedon, it became clear as one tried to piece the fragments of the skull together that this man had suffered from a slight facial deformity: it took practical experiments to confirm that these were not artefacts of cremation but represented his condition in life. Further, the crucial evidence of the eye wound was identified with certainty by fitting together the damaged parts of the skull so that they formed a working armature for the face where previous work had simply repaired them to what looked like a good fit (Bartsiokas quite fails to grasp this point in his recent study of the bones).[3]

It will be clear that although such 'manual' reconstructions may be laborious in terms of time and effort, just because of this the results are much truer and much less wooden than those created using a computer. For the present the artist's brain is still the more knowledgeable, experienced, and sensitive resource. When computer-based reconstruction was first introduced its practitioners tried to fit scanned images of the faces of living individuals of roughly the same physical type over an ancient skull: the clash this produced between the skull (the armature) and the features stretched over it was admirably demonstrated in a series of controlled experiments in Tennessee, still unpublished as far as we know.[4] The computer-based approach has developed and now uses an 'average' face of approximately the right age to mould over a particular skull. However, it still works on the basic principle of modifying an existing face rather than creating a new one from scratch, and it cannot fill in the unknown gaps any more than any other mechanical technique: while it can alter the shape and the dimensions of the scanned face it cannot alter its basic morphological features, such as the nose, mouth, and eyes. Aside from or perhaps because of this potential inaccuracy such an approach must produce a type rather than an individual, where the manual method relies on an individual skull to create a new face belonging to that individual. Although it did use the manual method, television's recent display of the reconstruction of 'a Jewish skull from the first century AD' under the headline 'Is this the face of Jesus?' reveals a basic misunderstanding of the individuality of each person's skull. The face which viewers were shown was that of an individual buried at Ein Gedi on the shores of the Dead Sea in a typical Jewish cemetery of the late Second Temple Period (*c*. 100 BC – AD 100) — in other words from the time of Christ, but the one thing that was certain was that it was not the face of Christ. Taking one step back, a study carried out in Italy some forty years ago already demonstrated the dangers of trying to assess physical appearance and even ethnic type from dry skulls without involving facial reconstruction.[5]

That the technique produces a recognisable likeness of an individual has been demonstrated many times by scientific controls carried out on cadavers in the Manchester Medical School and sometimes more publicly on living victims, as in the case of Professor Peter Egyedi, whose face was reconstructed 'blind' as a challenge from the Dutch National Association for Oral and Maxillo-Facial Surgeons and the Royal Belgian Association of Stomatology in 1996.[6] However, the most frequent and repeated controls are the forensic examples, well known to viewers of *Crimewatch UK*.[7] In these cases what matters is that the face should be distinctive enough to strike a spark of recognition in someone who knew the dead person. It then becomes interesting to ask just how and why we recognise a familiar face. Much has already been written about this question, notably by psychologists, and this is not the place to reopen that debate, save to note that there have been some intriguing cases where incorrect or insufficient information about the remains from the police surgeon has resulted in a reconstruction that was inaccurate in details such as age or race, but which has nevertheless provoked a recognition. When one compares the reconstruction with photographs of the individual in life in these cases, what they always share is their proportions, dictated by the skull beneath.

A striking example is provided by the 'Great Harwood case'. A body was discovered by the side of a road near Great Harwood in Lancashire very early in the morning of 19 March 1988, having been deliberately covered in petrol and set alight. Later examination showed that the dead person had been struck repeatedly around the head and chest, and there were the remains of flex around his neck. Despite the fire, which had obviously been intended to disfigure the body beyond recognition if not actually to destroy it, enough of the face remained to establish several details, such as the type of nose and mouth, and the skull beneath was not seriously damaged. It was suggested that there was no need for a reconstruction, and that even in this burned state it would be recognisable to those who knew the person in life, but it cannot be acceptable to show such charred and damaged remains to the public. A similar argument has been applied to the crushed and shrunken faces of bodies found in peat bogs, apparently wondrously preserved by the tannins in the peat. Though much less upsetting than the burned and lacerated victim from Great Harwood, such faces bear little resemblance to the way the person looked in life, for the chemical combination in the anaerobic environment of a peat bog, while preserving the soft tissues, also leaches out the inorganic components of the bones, softening them so that they can be distorted by the weight of the peat pressing on them from above. In both instances the argument overlooks the distortion which results from fire or immersion in peat, and the inescapable and rigid rules imposed by the skull in its undamaged state, which if properly applied allow no scope for artistic licence. In the Great Harwood case some of the information gleaned by the police surgeon under admittedly difficult circumstances proved to be simply incorrect. He deduced that the dead person (male) was in his early twenties and probably of Malay or Chinese race. The reconstruction which was shown on *Crimewatch* was of a young man of South-East Asian appearance. Within twenty minutes of the television broadcast he had been recognised by a member of

---

[3] Bartsiokas (2000).
[4] Addyman (1994).
[5] Heurgon (1964), 120–1; Wolstenholme and O'Connor (1959); summarised in Prag and Neave (1997), 185–6, with further references at 247 n. 10.

[6] Prag and Neave (1997), 228–30, plate XVI.
[7] Prag and Neave (1997), 33–9.

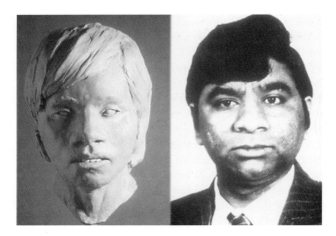

Figure 14.4 The reconstructed face of the Great Harwood victim, and a photograph of Sabir Kassim Kilu in life.

his family, living in the East Midlands, as Sabir Kassim Kilu, a middle-aged businessman of Indian race who had been murdered in a domestic feud. There are obvious differences in the angle and shape of the eyes and the epicanthic fold characteristic of East Asian peoples, for which the bone provides no evidence; however, these had not masked the underlying proportions of the face and the shape of the nose and the mouth (Figure 14.4).

If it can be demonstrated in this way that the technique works on modern, identifiable cases, then one can use it with some confidence to reconstruct faces on ancient skulls. Such reconstructions have become fashionable in museums in the last fifteen or twenty years, and we shall return to the visual and didactic aspects later. Intellectually the interesting and the challenging feature of facial reconstruction is that it can become another tool to be used in solving archaeological and historical problems.

Fundamental to any good reconstruction is a close study of the skeletal remains which can reveal details that identify the dead person. We have already referred to the damage on the skull from Vergina, overlooked or misunderstood by previous researchers, which our colleagues recognised as an eye wound and thus enabled us to identify it as that of Philip II. This reconstruction was first published as long ago as 1983, and was the first occasion when the technique in all its ramifications was used to answer a series of interlocking questions. First, who was buried in tomb II under the Great Mound in the village of Vergina in Macedonia? The archaeological evidence provided pointers that suggested this was a royal tomb, probably one built in the later fourth century BC, but it could not identify the dead. Having identified the one burial it became possible to suggest names for the other dead under the mound.[8] Second, having demonstrated that it was a royal tomb with a known occupant, then this in turn identified the modern village of Vergina with Aigai, the original capital of the Macedonians, where

they continued to bury their royal dead even after the capital was moved down to Pella, on the plain, *c.* 400 BC. This again had implications for identifying other burials and buildings on the site.[9] The identification has caused problems in the dating and comparative development of the arch and the vault, and in the introduction of certain Eastern themes, such as the royal lion hunt, into Western art, and it remains a source for debate. We do not wish to reopen that debate here, but we would argue that the traumatic damage to the skull taken together with other evidence, such as the construction history of the tomb and the circumstances of the burial, override the other considerations, and that tomb II at Vergina becomes one of those fixed points in archaeology from which stylistic dating of art and architecture must hang. Philip II was assassinated in 336 BC, and this becomes the *terminus ante quem* for anything found in it or built into it. In the same way no pottery found in the tomb of the Athenians at Marathon can be later than 490 BC, the date of the battle in which the men buried there met their death.

The fact that the technique produces an objective likeness of individual people suggested that it could be used to investigate kinship links through facial similarity in contexts where other evidence was lacking. The burials in Grave Circle B at Mycenae, dating from the sixteenth century BC and excavated by the Greek Archaeological Service between 1951 and 1954, appear to fall into four groups, in the north-west, north-east, and south of the circle respectively, with a final group comprising two graves a little to the west of the centre of the circle (Figure 14.5). Aside from an intrusive grave which does not concern us here, these two are among the latest if not actually the latest in the circle, but otherwise the three groups all span the three or four generations during which the grave circle was in use. The more famous and richer Grave Circle A, discovered by Schliemann in 1876 and now inside the walls and the Lion Gate,

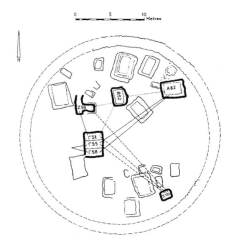

Figure 14.5 Grave Circle B at Mycenae: plan showing possible kinship links between occupants of the graves suggested by facial reconstruction.

[8] E.g. Andronicos (1984), 226–31; Musgrave (1990), 274–82. For a comprehensive list of publications where the possible identifications are discussed see Riginos (1994), 104 n. 1.

[9] Andronicos (1987); Tomlinson (1987).

overlaps with them and extends another generation or so downwards in time.[10]

Assuming that the grouping has real significance and is neither a creation of the archaeologists' imagination nor mere random chance, the problem is to try establish the relationships between the groups: are they branches of the same family, or different families? The significance of this lies in the fact that one group appears to win out over the others and establish itself in the wealthier, spectacular Circle A. Whoever those people were, in the mid-thirteenth century BC a later ruler of Mycenae — perhaps even the 'Agamemnon' who led the Greeks against Troy — felt they were so important that some 300 years after their final use he extended the walls of Mycenae and built the Lion Gate to take the graves of Circle A inside his city precinct, while Grave Circle B, lower down the hill, was ignored and forgotten. This is prehistory, so there are no written sources to help us, and in the mid-1980s it seemed worth testing whether facial reconstruction might provide an answer where other more traditional methods had failed.

The soil of the Argolid is very alkaline, and as a result the bones were very fragile. Further, except for the earliest burials, many of the graves had been reused, often more than once, and when the previous occupant's remains were pushed to one side to make way for a new burial it was clearly done with little care or ceremony. As a result, none of the skeletons from Circle A were now in a fit state to be reconstructed; in Circle B only seven skulls out of the thirty-five individuals buried in the twenty-six graves were fit to be reconstructed, and even those generally needed further consolidation and repair. More crucial, despite extensive searching, in only one case could the lower jaw still be located. As a rule this would have ruled out any valid reconstruction, since it is now clear that the form and set of the jaw have a very significant and individual influence on the structure of each face as a whole, even though this had not been established in 1986–7 when these reconstructions were made. However, fortunately J. L. Angel, the American anthropologist who studied the bones in 1954 when all the mandibles surviving from antiquity were still present, kept such good notes and photographs that it was possible to recreate the missing items in wax from his records.

A number of distinctive facial types emerged with the reconstructions. Skull 59 from the early grave Zeta (the numbering of the skeletons is Angel's; the Greek letters identifying the graves are those of the excavators, Papadimitriou and Mylonas), a battle-scarred and arthritic warrior in his late fifties, had a long face with high forehead, lantern jaw, and rather narrow features. His simple grave was one of the earliest in the circle and it seems plausible that he was one of the dynastic founders. His face shows strong similarities to skull 51 from grave

Gamma, another fighter, around twenty-eight years old when he died, who had survived other head injuries but apparently succumbed to a fractured skull whose effects trephination had failed to remedy. He was two or three generations younger, and had been given an unusually poor burial compared to those of his contemporaries (Figure 14.6). Surprising too was the way in which he had been buried across the feet of Gamma 55, a much wealthier man in this mid-thirties who was buried with an electrum mask (the only mask from Grave Circle B), several weapons, pottery, and other grave goods.

Grave Gamma contained at least four individuals. Skeletons 51 and 55 had been buried at the same time, and the legs of 55 had been drawn up to make room for the body of the younger man. When reconstructed Gamma 55 had a quite different type of face: heart-shaped, with wide cheekbones and wide-set eyes and rather delicate small features (Figure 14.7). These features appeared again on the face of Gamma 58, a woman who had been given a rich burial in the same grave a few months earlier (Figure 14.8). Her remains had been pushed to one side even before her body was completely decomposed and disarticulated to allow the two men to be interred. She too was in her mid-thirties when she died, suffering from the first twinges of arthritis in her lower back and her hands, a tall woman with the same heart-shaped, wide-cheeked face as the man 55, even if her features were not quite as delicate.

The remains of the other one or possibly two individuals from this grave were too fragmentary to allow a reconstruction, but those that we had been able to make suggested that in Grave Gamma the man 55 and the woman 58 had such similar faces that one had to assume that they were related: perhaps brother and sister or maybe cousins. They showed some similarity with Alpha 62 (Figure 14.9), a man of around twenty-three buried in the north-east part of the circle, but Gamma 51 and the early Zeta 59 from the north-west sector had quite a different type of face. This was not our imagination: other facial types had emerged too — for example Beta 52, from an early-middle-period grave which probably belonged with the north-eastern group, had a small head with prominent 'beaky' features which we did not find elsewhere (see Figure 14.6).

As the plan (Figure 14.5) shows, what makes this grouping of faces interesting is that grave Gamma is one of a pair of late graves, inserted in the centre of the circle towards the end of its period of use. Do the faces and the kinship they suggest imply some kind of dynastic or political rapprochement between the north-eastern and north-western 'families'? How does one explain the positioning of Gamma 51's remains and their relatively poor grave goods? Is this some kind of knight–squire relationship, perhaps?

We have touched on these questions in chapter 6 of *Making Faces* and in our 1995 article in *BSA* (see note 10), but they are for specialist Aegean prehistorians to answer: in this context our aim has been to develop and to exploit the potential of objective facial reconstruction as an additional tool in understanding the dynastic

[10] The excavation: Mylonas (1973), with Angel's discussion of the skeletal remains (in English) at 379–97. For a full account of these reconstructions, see Prag and Neave (1997), chapter 6 (105–45), and Musgrave et al. (1995).

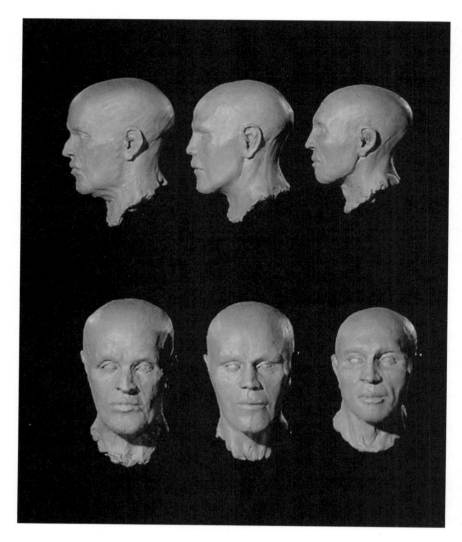

Figure 14.6 Reconstructed heads of Zeta 59, Gamma 51, and Beta 52 from Grave Circle B at Mycenae (profile and front views).

Figure 14.7 The face of Gamma 55 from Mycenae.

Figure 14.8 The face of Gamma 58 from Mycenae.

politics of Mycenaean Greece. It was interesting to compare our conclusions with those of Angel, who, working only from the dry bones but having the benefit of a vast experience of handling human remains from excavations in Greek lands, believed he could see evidence of kinship between a number of individuals from the

two grave circles.[11] He had no independent means for checking his suggestions, but his conclusions generally matched ours where they overlapped. However, we had

[11] Mylonas (1973), 389–90, with plates 245–6.

Figure 14.9 The face of Alpha 62 from Mycenae.

to restrict ourselves to those seven people whose skulls survived in a reasonably complete state, which meant that we had to take a fairly broad view of each individual, whereas Angel studied the entire population of the two circles, including their post-cranial bones, in very great detail, including the much more damaged and incomplete specimens, and felt confident enough to suggest relationships on what was sometimes very fragmentary evidence: for instance he proposed a kinship link between the beaky Beta 52 and individuals buried in the graves on either side (Delta 61 and Nu 66a) on the basis of the shape of the vaults of their skulls. Archaeologically such a link is perfectly possible, but of Delta 61's skull only the frontal bone survived even when Angel saw it, and Nu 66a was hardly more complete. Dr Angel showed great interest in our work, and gave us his enthusiastic support: we still regret that he died, in late 1987, before we could discuss our results properly with him.

It must be admitted that even of the seven skulls on which we worked some, such as Alpha 62, Gamma 55, and Gamma 58, were badly damaged, and detailed evidence for parts of the middle of the face around the nose and mouth were lacking. Forensic controls such as the Wyre Street case in Manchester in 1993–4 (the 'headless body case') and that of Harjit Singh Luther, whose remains were found in a field near Warsash in Hampshire with the skull broken in pieces by a plough, have demonstrated that even in such instances it is still possible to produce a valid and identifiable face, but it is probably true to say that in the Mycenae experiment we have pushed the reliability of facial reconstruction in the archaeological context to its limits.[12] However, unlike Angel we are now able to call on other techniques not only to control our work but also to extend it further. At the time when the Grave Circle B reconstructions were first planned in 1985 this could be seen as a radically new approach to such anthropological problems. Since then other biological or bioanthropological methods have

been developed to the point where they can usefully be applied to such questions. The two that are relevant to the Mycenaean kinship question are biological distance study and DNA analysis, neither of which is restricted to individuals with intact skulls.

Biological distance study (epigenetic variation) uses the small genetically linked variations in the form and structure of the skull and the teeth to trace familial and ethnic links. For some years Lisa Little at Indiana University has been using these genetic traits to seek out distinct biological units in the Aegean Bronze Age populations, with a particular focus on the intermarriage and the resultant mixing of groups at the end of the period: her research is now working in tandem with ours, as a methodological comparison. In the same spirit we are working with the Department of Biomolecular Science at UMIST on a project to seek out and identify the DNA of these peoples: Grave Circle B, then Grave Circle A, and the contemporary population from Lerna on the other side of the Argolic Plain. At a later stage we hope to move on to remains from other Aegean sites. Thus far DNA has been identified and sexed in nine out of twenty-two individuals from Circle B (no easy task, bearing in mind the heavy mineralisation of the bones caused by the very alkaline soil in which they have been buried).[13] The kinship analysis, along with comparison with bones from the other sites, is still in progress at the time of writing and it is too early to report. Ultimately, however, we should be able to check the results suggested by the facial reconstructions, and be able to extend them very much further. None the less, before one is tempted to dismiss facial reconstruction as a means of picking out kinsmen in favour of the 'modern science' of DNA analysis, it is worth bearing in mind that because of the fragmentary nature of the DNA this is very slow and painstaking work, and much the more costly in time and money of the two methods.

In the case of Philip of Macedon the fact that this kind of facial reconstruction entails a hands-on three-dimensional approach had helped to identify the all-important injury and malformation to the skull. In the case of the Mycenaeans it had a similar if not quite so crucial bonus. While it is much easier to pick out facial similarities on a series of bald, hairless reconstructions, they are not as satisfying nor indeed as complete historically as heads that have been given a full head of hair, even if the evidence here has to be secondary. In the case of the six men from Mycenae this was relatively simple to do, once one had decided which models to follow. The evidence from contemporary or near-contemporary objects, such as the electrum 'death mask' found with Gamma 55 and the gold masks from Circle A, along with the huntsmen illustrated on the famous daggers from Circle A, suggested that the men of this period wore their hair and beards in a variety of fashions, probably dependent in part on age and status. The reconstructions reflected this, and ranged from hair extending down over the nape and a full beard and moustache to short hair

---

[12] Wyre Street ('headless body case'): Prag and Neave (1997), 37–9; Warsash: Roy (1996), 13.

[13] Brown et al. (2000).

Figure 14.10 The coffin of Seianti in the British Museum.

and clean-shaven cheeks. To apply any of these styles to the clay heads presented no problem, as can be seen in Figures 14.7–9.[14]

However, when it came to the woman Gamma 58, the evidence was much less straightforward. The evidence[15] is really all two-dimensional, based on the small gold plaques found in Shaft Grave III in Circle A and on fresco paintings from Thera and Mycenae. From these it was clear that a Mycenaean lady of the Middle–Late Bronze Age wore her hair in an elaborate style which could include a row of tight curls over the forehead and two or three spiky curls or plaits in a row along the top of the head from the crown forward to the brow: the foremost of these might hang down over the forehead, but the others seemed to be held in place with some kind of pomade. The back hair was pulled tight over the crown of the head and allowed to hang down in three long locks or as a single pony-tail. It was hard enough to understand this fashion in the two-dimensional working drawings kindly provided by Diana Wardle of Birmingham University; modelling it in three dimensions added a new dimension to the understanding and interpretation as well, and after a false start it was clear that the only way to make the Mycenaean hairstyle 'work' was to make a bust rather than just the usual head. Otherwise the hair would not hang properly (Figure 14.8).

Another area where the individuality of this type of facial reconstruction can serve as a useful research tool lies in the study of realistic portraiture. It is often suggested that realistic portraiture in the West begins in Etruria, and the refurbishment of the British Museum's gallery on 'Italy before the Romans' provided an opportunity to test this, when Dr Judith Swaddling proposed a reconstruction of the face of Seianti Hanunia Tlesnasa, an

Etruscan noblewoman who died *c.* 200 BC. Her terracotta coffin (Figure 14.10) containing the almost intact skeleton was discovered near Chiusi in 1886 and acquired by the Museum (GRA 1887.4–2.1–2) in the following year. Her name had been cut into the base while the clay was still wet, showing that the coffin had been made to order. On the lid reclines a life-sized figure of a rather plump woman, finely dressed with jewellery on her arms, neck, and ears and a tiara on her head, holding back her cloak in the traditional gesture of a bride revealing her face to her new husband. There is no reason to doubt that bones and coffin belong together, and this seemed an ideal opportunity to check on the faithfulness or otherwise of the image on the lid.

There is a long tradition of faces with apparently idiosyncratic features on the lids of Etruscan cinerary urns and coffins, quite unlike the idealising traditions of Greek art, but it is only around 350 BC that a sea change overtakes them. Although it is clear that the Etruscans had always been interested in the particular person rather than the ideal, before this date the usual artistic practice was to add individual characteristics to typical faces.

There appears to be a serious possibility, not to be disregarded, that a turn from the 'typical' to 'real' portraits indeed happened in Etruria about or shortly after 350. If the workshops of the sculptured sarcophagi led this change, they merely brought the old Etruscan insistence on facial differentiation as a mark of the human reality to its ultimate conclusion. Henceforward portraits in the modern sense, that is, genuine likenesses of specific and nameable persons, may be expected in Etruria, though we will have to regard them as special and perhaps exceptional.[16]

In other words this might be the start of true portraiture in Western art, a practice that fulfils two basic requirements: it sets out to create a specific likeness,

---

[14] The others are illustrated in Prag and Neave (1997), 131–9, figures 11–25, and Musgrave et al. (1995), 122–5, plates 17–18.
[15] Discussed in detail in Prag and Neave (1997), 137–9, and Musgrave et al. (1995), 122–5.

[16] Brendel (1995), 396.

Figure 14.11 (a) The skull of Seianti; (b) her face reconstructed as an old woman; (c) the 'portrait' on her coffin.

and that likeness is of a named and nameable person.[17] However, Brendel pointed out — and his book was originally published in 1978 — that there seemed to be no way in which one could actually control 'the unanswerable question of personal similitude. The subjects of this stony portrait gallery are beyond our reach. In no circumstance can the portrait likeness of their image be taken for granted.'

For an artist who starts from a known personality that likeness needs to include something of the sitter's character. By contrast, facial reconstruction comes to the person's physical remains 'cold', without knowledge of the personality, but it will produce an essential likeness based solely on the physical evidence. The problem in Seianti's case was that the first analysis of that physical evidence carried out by Professor Marshall Becker suggested that these were the remains of a very old woman, aged between eighty and ninety, and this of course was what the reconstruction had to show. If that was the case, the figure on the sarcophagus was certainly not a portrait of Seianti (Figure 14.11).

This age estimate raised a number of other problems, discussed elsewhere.[18] Not the least of these concerned the process of making the sarcophagus. This is a large and technically elaborate piece, which cannot have been made in the short interval between Seianti's death and the time when her remains must needs be properly interred, bearing in mind that all this happened in a Mediterranean climate. And yet the inscription proves beyond doubt that it was made to order. These elaborate Etruscan sarcophagi, in many ways the equivalent of carved or modelled Greek grave-markers, were not intended for display above ground, but once occupied were placed underground in a burial vault either singly or with other members of the family. However, there is no evidence that Seianti might have had her coffin made in advance and kept it by her as some kind of *memento mori*, particularly given the size of the ensemble, although as we shall see an argument *ex silentio* is of doubtful validity in this case.

The case for Seianti's advanced years was based partly on the appalling state of her teeth, partly on the evidence of acute erosion of the pubic symphyses in

her pelvis. Such great age is not in fact as uncommon as might be expected among the Etruscan population. However, there was other evidence, such as the extent of the closure of the sutures in her skull, but above all the lack of geriatric degeneration of the spinal column or any signs of osteoporosis that caused the pathologist and the anatomist in our team (Dr R. W. Stoddart, Department of Pathological Sciences, University of Manchester, and Dr J. H. Musgrave, now at the School of Veterinary Science, University of Bristol) to believe that she was only middle-aged, a woman who had borne at least one child but who at the time of her death had not long passed the menopause. Therefore her age was tested by measuring the ratios of the two forms of dentine in two of her teeth (sclerotic apical dentine compared with non-sclerotic coronal dentine), work carried out by Dr David Whittaker (Dental School, University of Wales College of Medicine, Cardiff) and later retested by Dr Whittaker's colleague Dr G. J. Thomas. This is a very accurate technique, developed by Dr Whittaker and tested on the skeletons from the Spitalfields cemetery in London, where the parish records give a full control of the ages of the individuals interred there. Whittaker and Thomas agreed on an age at death for Seianti of around fifty, perhaps as young as forty-six and certainly no older than fifty-five.

It then became necessary to explain the fact that Seianti had lost all her molars and premolars, except for those in the left upper jaw, long before her death, that she suffered from acute arthritis in the jaw (particularly her right temporo-mandibular joint) which would have made chewing and speaking painful and caused her jaw to lock from time to time, and that to walk or ride would have given her much pain because of the arthritis in her right hip. From his study of the bones Dr Stoddart proposed the following scenario: in her youth Seianti had been an enthusiastic horsewoman, as the powerful muscle-attachments on her thigh-bones demonstrated. In her late teens she suffered a serious riding accident when her horse probably fell onto her, wrenching her right leg backwards and giving her a savage blow on the right side of her face. No bones were broken, but ligaments and probably muscles and tendons in her right leg and pelvis area were torn and there was internal bleeding into the soft tissues and joints. Although Seianti resumed her equestrian activities, within a year they became too painful, for the residues of the blood calcified and formed osteophytes, small bony leaflets or growths which if they grow in the path of a nerve can trap it, or on a joint can lead to arthritis. This happened in her right leg and hip, and in her jaw, where if it did not knock her teeth out the blow at least loosened them so that she lost them prematurely.

Dr Stoddart had one further deduction to make, this time from the uneven staining of the bones, the fact that some of the small bones in the hands and feet were missing along with the sternum, and the corresponding lack of any staining on the floor of the coffin which might have derived from decomposition of the body. From these it seems highly probable that her body was allowed

[17] E.g. Brendel (1995), 392ff., with further references at n. 19.

[18] Prag and Neave (1997), 177–85; Swaddling and Prag (2002).

(or assisted) to decay until it was reduced to a skeleton *before* it was placed in the coffin. This is not the place to set out his arguments, which — together with the rest of the story of Seianti — appeared in summary form in chapter 9 of *Making Faces* (especially pp. 178–85) and were published in detail in the monograph on Seianti.[19] For the present context they are important because they make the state of Seianti's skeleton compatible with the proposed age of around fifty, and they also explain how the artist could make this large and complicated coffin between the time of her death and her eventual burial. If the body was being excarnated in the undertaker's workshop, he had sufficient time to make such an elaborate item with its 'portrait', and certainly had opportunities to make sketches and models of her face.

The revised reconstruction shows a middle-aged lady, rather plump because this was what the medical evidence suggested. At first sight it still does not look quite a portrait, but just as in the case of Sabir Kassim Kilu described above, comparison of the underlying proportions, which are what give every face its distinctive structure and identity, shows that they are the same, with one small difference: Figure 14.12 illustrates the reconstruction and the face on the coffin matched by the forensic technique of photocomparison ('face mapping'), used by the police to compare images from security cameras with

photographs of suspects they have arrested. In Seianti's case the spatial relationship between the morphological features of the two faces (notably eyes, nose, and mouth) is very close, as is the distance separating the inner canthus of the eyes and the width over the wings of the nose (dictated by the piriform aperture in the skull beneath). What distracts the eye is that the artist, working at a period when Etruscan art had come under the idealising and even prettifying influence of Hellenistic Greece, has allowed himself to modify his rendering of Seianti's face in a few small particulars in order to make her more attractive. He has 'improved' two features in particular, and he has done this partly by altering the relative proportions of the upper and lower face (the distance between the bridge of the nose and the nasio-labial junction where the nose and upper lip meet, and between this junction and the point of the chin), making the lower part smaller than it was in real life (and on the reconstruction). This has given her a younger, more childlike face, with a more feminine 'little girl's' mouth. At the same time he has reduced the hollow at the bridge of her nose to give her a much sought-after 'classical' profile, and thus made her eyes more deep-set and a little larger and more soulful. It is important to see that the differences are much less noticeable from the front than they are from the side: a visit to the 'Italy before the Romans' gallery in the British Museum will immediately make it clear that, whether destined to be hidden underground or not, the whole conception of the sarcophagus is a frontal one. Allowing for these aesthetic modifications the reconstruction and the portrait have far too much in common to suggest that the artist was *not* intending a realistic and recognisable image.

The next step will be to test this conclusion against other possible tomb portraits — not such a simple matter, for despite what we have just said the old Etruscan tradition of customising typical faces by the addition of conventional individualising traits, such as furrowed brows or straight noses, persisted well into the Roman period, and probably only a small proportion of Etruscan artists or their customers were concerned to commission a truly realistic image.[20] Second, although museums in Italy and elsewhere may appear well stocked with Etruscan sarcophagi, in only a small proportion do the bones survive too. Finally, one has to remember that through most of their history the Etruscans preferred cremation to inhumation. Intact Etruscan skulls surviving along with their 'portraits' are not so easy to find.

Among the best-known examples of seemingly realistic portraiture in antiquity are the likenesses set over the faces of mummies from Roman Egypt, painted either on wooden panels placed over the wrappings or on linen shrouds stretched over the face of the deceased. Even a cursory glance through a series of these images shows a range of different facial types and idiosyncrasies that suggests that the artists were intending to depict individuals, although just as in Etruria there is a penchant for long straight noses and large deep-set eyes that must

Figure 14.12 Photocomparison of the reconstructed face of Seianti and the 'portrait' on her coffin.

---

[19] Swaddling and Prag (2002), 35–7.

[20] Barker and Rasmussen (1998/2000), 291.

owe more to artistic convention or fashion than to reality, and these features must be discounted in one's search for individual people. In some instances portraits that suggest family resemblances have on further study proved to have come from the same tomb and to have belonged to members of the same family—though it is of course hard to prove that the facial features such as cleft chins that are interpreted as family traits were not stylistic quirks of the painter.[21]

One approach to these mummies and their portraits is described by Linney et al. in Chapter 15, but for the sake of completeness it is worth mentioning here three reconstructions carried out by the Manchester Unit of Art in Medicine in the late nineties, in part as a control of this realism. In early 1997 two faces from the second half of the first century AD were reconstructed for the exhibition 'Ancient Faces' at the British Museum, one of a man and one of a woman, both found by Sir Flinders Petrie at Hawara.[22] They were of course done 'blind', and only when the work was complete were the results compared with the portrait panels: the similarity was quite uncanny, and there could be no doubt that these were the same people, and that with only the smallest of 'improvements' such as the characteristic enlarging of the eyes these were faithful portraits. 'Bruiser', as he became known (for neither name has survived), had as fleshy and aggressive a face in real life as on his portrait (Figure 14.13), while 'Fatima' had the same rather low forehead and prominent pointed chin on the portrait and in 'life' (Figure 14.14).

The question of when the portraits were painted—during the deceased's lifetime or immediately after death—and indeed whether they do faithfully represent the person on whose mummy they have been placed has long been a matter of debate. Radiological studies have generally confirmed the match, and CT scans of all the complete mummies in the 'Ancient Faces' exhibition revealed a consistent match of age and sex between mummy and portrait.

Figure 14.14 Portrait panel and reconstruction of mummy EA 74713 (British Museum).

However, there are exceptions. It can be argued that panels could be painted in advance and simply cut to size and incorporated into the mummy when the time came, but it would be more difficult to apply this line of reasoning to the portraits of children and others who died prematurely. Indeed, X-ray work has yielded several examples where there is total disagreement in age and/or sex between the appearance of the portrait and the evidence of the physical remains inside the wrappings.[23] The fact that in some cases even the sex is wrong gives the lie to any notion that when middle-aged faces accompany the mummies of very old men the portrait might have been painted during the deceased's lifetime.

The portraits of both the British Museum mummies were painted on panels. At much the same time we reconstructed the face of a mummy of *c.* AD 50, also found by Petrie at Hawara but now in the Ny Carlsberg Glyptotek in Copenhagen.[24] Because the mummy began to 'sweat' after its arrival in Copenhagen in 1911–12 and there were fears that it might be beginning to disintegrate, the wrappings were cut open along each side so that the entire front could be removed in one piece for conservation. In fact the body inside was well preserved, and while it was free of its wrappings the dried and desiccated face was photographed. Professor C. C. Hansen, Professor of Anatomy at Copenhagen University, examined the face alongside the portrait, and concluded that the portrait was indeed a close likeness of the dead man: he published a photomontage combining the two to support his case.[25] Although we have already warned of the dangers of trying to assess the appearance of the face from a dry skull, or even a face that retains some dried tissue, it none the less came as a surprise to find that the Manchester reconstruction of the Copenhagen mummy bore very little relation to the portrait, which shows a

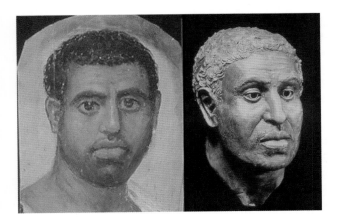

Figure 14.13 Portrait panel and reconstruction of mummy EA 74718 (British Museum).

21 Borg (1996), 98; Walker and Bierbrier (1997), 14–15.
22 Walker and Bierbrier (1997), nos. 18 and 23, acc. nos. EA 74718 and EA 74713; Prag (2002). See also Douglas (2001).
23 E.g. Filer (1997), 121–2. See Wilkinson et al. (2003) for a more extensive study of the problems of the 'match' between portrait and mummy, including those discussed here.
24 Acc. no. Æ IN 1425: Jørgensen (2001), 326, no. 32, and 356, no. 43, with further references.
25 Jørgensen (1998), 68–70, figures 17–19; *id.* (2001), 39–42.

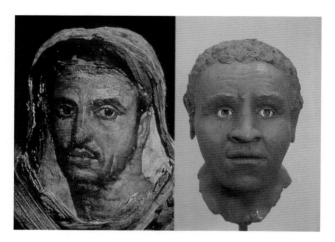

Figure 14.15 Portrait panel and reconstruction of mummy Æ IN 1425 (Ny Carlsberg Glyptotek, Copenhagen).

man perhaps in his thirties, with a rather long, severe face narrowing down to a small mouth and chin, and a fairly long nose and high forehead. By contrast the reconstruction has a square face with a wide mouth and nose that at first sight look almost negroid, yet with close-set eyes and a sloping forehead; the proportions of the face, particularly around the centre, are far more compressed (Figure 14.15). It is quite impossible that these represent the same man, or even that the artist has taken liberties in painting the portrait to modify his sitter's appearance: medical examination showed the remains to be those of a middle-aged man, not yet fifty, but the face really does not fit.[26] This is even more surprising when one realises that when a computer-generated reconstruction was made of the skull of 'Hermione', found together with the Copenhagen mummy by Petrie at Hawara in 1911, the portrait was proved to be a faithful likeness.[27]

On the Copenhagen mummy the portrait was painted on the linen shroud, which raises the question of whether it was done before the cloth was stretched over the mummy, giving the painter a chance to work if not from life at least direct from the face of the dead man, or whether he was simply called in to paint a face over a ready-wrapped mummy. In favour of the former is the fact that the edges of the portrait are concealed by the final layers of wrappings, applied to hold everything in place, something which often happened to the painted panels too. In that case one has to assume that the embalmers muddled up two bodies: since for the mourners the portrait was visible testimony as to the identity of the dead man, at least two families must have ended up with the wrong body. On the other hand the painting displays none of the tiny cracks which one might expect from the manhandling as the cloth was stretched into place. Perhaps the painter's instructions were none too clear, and the problem is one of lack of interdepartmental

communication at the undertakers rather than straight malpractice.

We have tried to summarise the intellectual content and purpose of facial reconstruction as a research tool. However — particularly in the twenty-first century — such an academic approach cannot stand in effective isolation from the 'real' world. Aside from the obvious forensic and medical applications, what other 'benefits' might be seen to derive from such detailed, painstaking, and (even after a quarter of a century) still exciting work?

For the archaeologist, the discipline of the three-dimensional approach and the concomitant value of the three-dimensional report are immensely stimulating. From the layman's point of view, the most basic product may simply be to produce a 'real person' for a museum display, a 'real Roman' in a gallery about Roman life, a 'real Viking' in the story of a Dark Age town. In many ways this is but a step up from the old-fashioned diorama that used simple mannequins to represent anthropological or historical types. It can be argued that here the computer-driven reconstructions come into their own. In terms of time and manpower they are quick and cheap when compared with the laborious and painstaking 'manual' method, and often the idiosyncratic detail that marks out an individual identity is of less importance.

Whatever method is used, to place a reconstructed human head in a museum display raises interesting questions about the viewer's reaction — a realistic waxwork may simply be too dramatic, and another medium, such as bronzed resin, is often more suitable for conveying the information which one wishes to impart. It is virtually impossible to discuss this in print or in a lecture illustrated with slides, because by their very nature these miss the point and the impact of the third dimension. On the screen or on the page a well-made waxwork surely conveys a very powerful and realistic impact. However, part of the *frisson* of meeting a murderer or a formidable politician in Madame Tussaud's comes from the shock of the realism of the figure before one — 'you'd think he was alive, standing there in front of me'. In a museum setting this is probably less appropriate unless one is intending to provoke the viewer into thinking about the moral and ethical questions of working with human remains, something which we have discussed elsewhere and which is becoming an increasingly important topic in archaeological and museum ethics. The current debate in Israel is perhaps the most extreme example, but the whole issue is now a sensitive one.[28]

Note

As well as those colleagues already named in the text (most of whom have borne with our sometimes strange requests for assistance over at least twenty years), the authors would like to thank Drs Mogens Jørgensen, Anne Marie Nielsen, Judith Swaddling, John Taylor, and

[26] Jørgensen (2001), 356; Prag (2002).
[27] Walker and Bierbrier (1997), 15–16, 37, no. 11, and 215. The reconstruction does not appear in the publication, and we owe this information to Dr Walker.

[28] Prag and Neave (1997), 219–27. There is a large, and growing, bibliography on the subject: see e.g. Walker (2000).

Susan Walker for their help in providing information and illustrations, and Dr Ingrid Strøm for help with a translation from the Danish. For permission to reproduce illustrations used in the article, we should like to record our thanks to the British Museum, the Ny Carlsberg Glyptotek, Lancashire Constabulary, the South Wales Police, and the Unit of Art in Medicine and the Manchester Museum in the University of Manchester.

References

ADDYMAN, S. M. (1994), 'The use of three-dimensional laser imaging for facial reconstruction', unpublished MA thesis (Knoxville, TN).

ANDRONICOS, M. (1984), *Vergina: the Royal Tombs and the Ancient City* (Athens).

ANDRONICOS, M. (1987), 'Some reflections on the Macedonian tombs', *BSA* 82: 1–16.

BARKER, G. and RASMUSSEN, T. (1998/2000), *The Etruscans* (Oxford).

BARTSIOKAS, A. (2000), 'The eye injury of King Philip II and the skeletal evidence from Royal Tomb II at Vergina', *Science* 288, 2 April: 511–14.

BORG, B. (1996), *Mummienporträts* (Mainz).

BRENDEL, O. J. (1995), *Etruscan Art*, 2nd edn. (London and New Haven).

BROWN, T. A., BROWN, K. A., FLAHERTY, C. E., LITTLE, L. M. and PRAG, A. J. N. W. (2000), 'DNA analysis of bones from Grave Circle B at Mycenae: a first report', *BSA* 95: 115–19.

DOUGLAS, K. (2001), 'Image is everything', *New Scientist*, 8 December: 39–41.

FILER, J. (1997), 'If the face fits...: a comparison of mummies and their accompanying portraits using computerised axial tomography', in M. Bierbrier (ed.), *Portraits and Masks: Burial Customs in Roman Egypt* (London): 121–6.

GERASIMOV, M. M. (1971), *The Face Finder* (London and Philadelphia).

HEURGON, J. (1964), *Daily Life of the Etruscans*, 2nd edn. (London).

JØRGENSEN, M. (1998), 'Menneskelige og guddommelige ansigter fra det gamle Ægypten', in A. M. Nielsen (ed.), *Det Sidste Ansigt* (Copenhagen): 53–71.

JØRGENSEN, M. (2001), *Ny Carlsberg Glyptotek: Catalogue Egypt III. Coffins, Mummy Adornments and Mummies ... 1080 BC – AD 400* (Copenhagen).

MUSGRAVE, J. H. (1990), 'Dust and damn'd oblivion: a study of cremation in ancient Greece', *BSA* 85: 271–99.

MUSGRAVE, J. H., NEAVE, R. A. H. and PRAG, A. J. N. W. (1995), 'Seven faces from Grave Circle B at Mycenae', *BSA* 90: 107–36.

MYLONAS, G. E. (1973), *O Taphikos Kyklos B ton Mykenon (Grave Circle B at Mycenae)* (Athens).

PRAG, A. J. N. W. (2002), 'Proportion and personality in the Fayum portraits', *BMSAES* 3: 55–63 (http://www.thebritishmuseum.ac.uk/bmsaes/issue3/prag.html).

PRAG, J. and NEAVE, R. (1997), *Making Faces Using Forensic and Archaeological Evidence* (London: repr. with corrections 1999).

RIGINOS, A. S. (1994), 'The wounding of Philip II of Macedon', *Journal of Hellenic Studies* 114: 103–19.

ROY, A. (1996), 'Shallow grave facts are stranger than fiction', *The Daily Telegraph*, 31 July 1996 (science section): 13.

SWADDLING, J. and PRAG, A. J. N. W., eds. (2002), *Seianti Hanunia Tlesnasa: the Story of an Etruscan Noblewoman*, British Museum Occasional Paper 100 (London).

TOMLINSON, R. A. (1987), 'The architectural context of the Macedonian vaulted tombs', *BSA* 82: 305–12.

WALKER, P. L. (2000), 'Bioarchaeological ethics: a historical perspective on the value of human remains', in M. A. Katzenberg and S. R. Saunders, *Biological Anthropology of the Human Skeleton* (New York): 3–39.

WALKER, S. and BIERBRIER, M. (1997), *Ancient Faces: Mummy Portraits from Roman Egypt* (London).

WILKINSON, C., BRIER, B., NEAVE, R. and SMITH, D. (2003), 'The facial reconstruction of Egyptian mummies and comparison with the Fayuum portraits', in N. Lynnerup, C. Andreasen and J. Berglund (eds.), *Mummies in a New Millen[n]ium: Proceedings of the 4th World Congress on Mummy Studies, Nuuk, Greenland, September 4th to 10th, 2001*, Danish Polar Center Publication no. 11 (Copenhagen): 141–6.

WOLSTENHOLME, G. E. W. and O'CONNOR, C. M., eds. (1959), *A CIBA Foundation Symposium on Medical Biology and Etruscan Origins (London 1958)* (London).

CHAPTER 15

Reconstruction of a 3D Mummy Portrait from Roman Egypt

Alf Linney, João Campos & Ghassan Alusi

Introduction

The subject of the reconstruction described here, the portrait mummy of Hermione, was excavated by Flinders Petrie at Hawara in 1911 and was finally presented to Girton College, Cambridge, after being purchased from the British School of Archaeology at a cost of £20. Hermione lived during the reign of the Roman emperor Tiberius in the Fayum. Greek immigrants to ancient Egypt were sometimes mummified. Although somewhat neglected by historians, it is thought that this group of people were the descendants of soldiers who fought with Alexander the Great and the Ptolomies. They apparently adopted Roman fashion, had a Greek social and cultural identity, and believed in the Egyptian gods who offered a firm prospect of eternal life. Hermione is believed to have been a schoolteacher, as her coffin portrait bears the Greek inscription 'Hermione grammatike'. The three-dimensional reconstruction of Hermione described here was carried out for the exhibition held at the British Museum '"Ancient Faces": Mummy Portraits from Roman Egypt' (14 March–20 July 1997), which was presented in association with the Fondazione Memmo of Rome. It was later shown at the Palazzo Ruspoli in Rome from October 1997 to early 1998.

X-ray Imaging and Mummies

The use of X-rays for the study of mummy anatomy is not new. Very shortly after Roentgen's discovery of X-rays in 1895 archaeologists started to use the penetrating power of this new form of radiation to produce shadowgraphs showing what was hidden inside well-wrapped ancient relics. The first report to include X-ray studies on mummy material was by the well-known Egyptologist Petrie, and appeared in 1898.[1] One of the figures in this report shows an X-ray view of legs of dissevered mummified bodies in wrappings.

In recent years X-ray imaging techniques have been used by a number of research groups to examine Egyptian mummies. Considerable information has been gained without removing the painted cartonnage and the wrappings protecting the mummy.

As well as the non-invasive nature of these techniques, they can provide invaluable information on the

medical state of the body at the time of death and other information relating to the person's life, as well as the process of mummification itself. An atlas of mummy X-rays was published in 1980.[2]

Over the past thirty-five years, papers have been published on the use of film radiography,[3] xeroradiography,[4] and, since its introduction in the 1970s, computer tomography (CT) for the study of mummies.

Topically, X-rays of the famous Pharaoh Tutankhamen taken in 1969 and again in 1978, which show a thickening of the cranium, are still being used to raise questions about how he died. Was it a natural death or was he murdered?[5]

Computer tomography is an X-ray medical imaging method, in everyday use in hospitals, which has been increasingly used in the study of mummies. The technique has the advantage of providing a measure of the three-dimensional distribution of X-ray absorption coefficients throughout a scanned volume. Data in this form may be used to create a three-dimensional reconstruction of the volume, which can then be visualised or even reproduced as a solid object. Three-dimensional computer tomography is a useful method of non-destructively evaluating palaeopathological remains, and provides information which cannot be obtained by any other means. CT scanning has now been used for more than two decades to study ancient mummies. One very early study using this imaging technique looked particularly at the cranial cavity[6] and others soon followed.[7]

Hermione's mummy had previously been X-rayed in the Research Laboratory of the British Museum in July 1952. This examination revealed that the mummy package contains the complete remains of a human being.[8] Although it is possible to understand the anatomy of a mummy from such an X-ray shadowgraph, it is not possible to reconstruct a three-dimensional model of the anatomy. The X-ray study of Hermione's mummy did, however, reveal that she was a delicately boned

---

[1] Petrie (1898).

[2] Harris and Went (1980).

[3] Gray and Dawson (1968).

[4] Middleton et al. (1992).

[5] Brier (1998).

[6] Lewin and Harwood-Nash (1977).

[7] Harwood-Nash (1979); Vahey and Brown (1984).

[8] Filer (1997).

woman and that she died somewhere between the ages of sixteen and twenty-five.

For the three-dimensional reconstruction of the face of Hermione, four procedures were necessary: the acquisition of three-dimensional data on what lies inside the wrappings, the three-dimensional reconstruction of the skull, the reconstruction of the soft tissue over the skull, and finally the application of texture to the reconstructed facial surface.

Acquisition of Data

CT imaging systems consist of a couch on which the subject is placed which advances under computer control into a rotating or spiralling X-ray scanning beam. The absorption of X-rays is measured across the scan and a computer converts the absorption data into a digital X-ray image of a cross-sectional 'slice' of the body. In these X-ray images dense materials, such as bone, appear white; less dense materials, such as skin or cloth, appear grey; and air appears black. As the couch moves through the scanner, the body is effectively scanned in slices producing sequential cross-sectional images. These images may be used directly to provide information about such details as the age and sex of the mummy, and the general state of health. Sometimes evidence can be elicited on the cause of death and the nature of mummification. Hidden artefacts are often also detected within the wrappings.

Hermione's mummy was scanned using an ELSCINT high-resolution X-ray CT scanner which is used for medical imaging in diagnosis and surgical planning in many hospitals. Computer tomography allows cross-sectional imaging of anthropological as well as clinical subjects. A scan resolution of 512 × 512 pixels was used: 600 slices were obtained at 0.5 mm spacing on the head and 260 slices for the rest of the body at 5 mm spacing. The individual slice width was 1.2 mm. Figure 15.1 shows the scanning of Hermione in progress.

This multiple slice data was then directly visualised as a volume using Medica3D software developed by MTT AG, Fribourg, Switzerland, for medical applications. The software was run on a Silicon Graphics Max Impact Indigo 2 workstation, and images were produced on the screen by a technique known as volume rendering.[9] The technique for producing images on a video monitor uses optical cues to create the impression of a three-dimensional anatomical object. The perception of three dimensions is further enhanced by rotation of the image at a suitable speed to produce parallax cues. The method of visualisation or rendering also allows various parts of the anatomy to be displayed as opaque or semi-transparent surfaces, giving a full appreciation of the relationship between structure at different depths within the anatomy. The software used also facilitates multi-planar reformatting of the data, so that any anatomical section can be selected for display, allowing for further

[9] Linney and Alusi (1998).

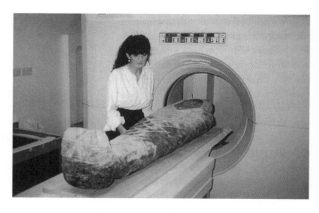

Figure 15.1 The scanning of Hermione in progress.

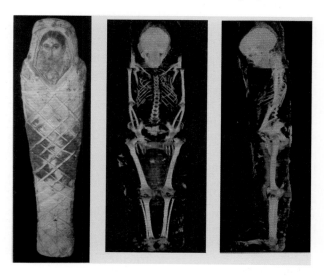

Figure 15.2 The mummy with 3D visualisations of the whole skeleton in anterior and left lateral views.

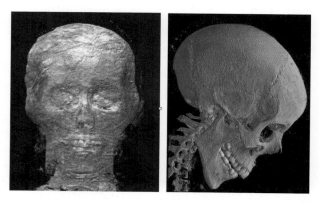

Figure 15.3 The mummified face and volume-rendered skull of Hermione.

detailed exploration of the anatomy. The mummy together with 3D visualisations of the whole skeleton in anterior and left lateral views is shown in Figure 15.2.

By adjusting the parameters controlling tissue transparency, images of facial features were produced, as if the mummy had been unwrapped. The face of the mummy is shown in Figure 15.3. An image of the skull stripped of mummified tissue is also shown in this figure. Three-dimensional reconstruction provided images of facial features as if the mummy had been unwrapped.

A number of other groups have reported on the use of CT and 3D data visualisation to produce images of the anatomy of a mummy.[10]

Reconstruction of the Three-dimensional Facial Surface

The reconstruction of soft tissues over skulls has been practised for centuries. In modern times facial reconstruction is used in forensic anthropology to aid in the identification of skeletal remains. The traditional method is to apply clay or similar modelling materials over the skull or cast model to a depth which is consistent with the thickness of soft tissue expected to be found in the living person. This depth depends very much on the position on the skull surface and also on the background of the subject, such factors as gender, ethnic group, and lifestyle being important. The actual procedure uses known measurements of the thickness of soft tissue over key areas on the human face. A cast of the skull is made and wooden pegs cut to lengths of the desired tissue thickness are attached to the corresponding points of the skull.

Layers of clay are built up to the height of pegs and final touches are made to model the skin surface. The nose, lips, and eyelids are modelled on top, their shape determined by knowledge of the underlying muscles.[11] The procedure may also involve a greater use of anatomical knowledge to construct the actual components of the facial anatomy: for example, building up muscle groups onto which is developed the facial surface to provide added realism. The methodologies and results have been described in detail.[12] The tissue thicknesses used in such reconstructions have been collected by physical measurement on cadavers for the purposes of forensic reconstructions.[13] CT imaging has recently been used,[14] and more recently MRI (magnetic resonance imaging) and ultrasound imaging for this purpose.[15] MRI and ultrasound have the advantage that they may be used on living subjects.

The approach we have adopted is an electronic analogue of the clay modelling method. For the reconstruction of the facial surface of Hermione, tables of tissue thickness[16] have been used, combined with the warping of a 3D model of a facial surface. The data for the facial model was collected using an optical surface scanner developed for clinical purposes at University College, London.[17] A schematic diagram of the scanning system is shown in Figure 15.4. The technique is to record online the shape of a line of light projected onto the face. The

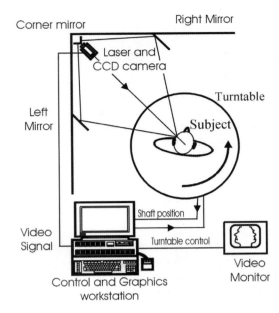

Figure 15.4 The facial scanning system used at University College, London.

line is viewed obliquely by a CCD video camera via two mirrors either side of the face. This viewing system eliminates the loss of data due to shadowing of parts of the face by prominent features, principally the nose. Simple geometric analysis using the principles of triangulation allows this shape to be transformed into a set of three-dimensional co-ordinates of points along the line illuminating the facial surface. By rotating the subject under computer control, the whole of the facial surface is scanned, and approximately 50,000 co-ordinate measurements are made on the facial surface in approximately 7 seconds, with an accuracy of 0.5 mm.

In reconstructing a face over a skull the choice is between using an average facial surface derived from a collection of data on a specified population of individuals or a carefully chosen individual face. Both average and individual faces were tried, and it was decided that it was more appropriate to use data from the average female face for the reconstruction of Hermione's face.

The choice of the average female face was based on the appearance of Hermione's skull as visualised and our experience of which group of facial shapes would be most consistent with this. For the reconstruction of Hermione's face, 54 tissue thickness reference points corresponding to recognised anatomical landmarks were used. The points on the skull and the corresponding points on the facial surface were placed interactively by use of a mouse-driven cursor. The warping of the facial surface to the mummy skull is illustrated in Figure 15.5. Figure 15.6 shows the 3D reconstructed face of Hermione from two viewpoints.

The method of deforming the facial surface to the mummy skull, based on spline interpolation between the 54 reference points, was developed more than a decade ago by our group when a comparison was made with the results of those produced by the traditional clay

[10] Marx and D'Auria (1988); Pickering et al. (1990); Isherwood and Hart (1992); Baldock et al. (1994).

[11] Neave (1979).

[12] Prag and Neave (1997).

[13] Suzuki (1948); Rhine and Campbell (1980); Rhine et al. (1982).

[14] Phillips and Smuts (1996).

[15] Manhein et al. (2000).

[16] Rhine et al. (1982).

[17] Linney et al. (1997).

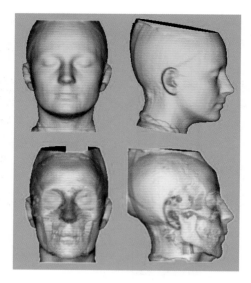

Figure 15.5 Average female facial surface warped to the mummy skull.

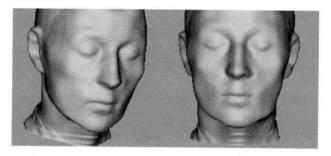

Figure 15.6 Reconstructed facial surface of Hermione.

modelling method.[18] Although similar results are produced, the computer method does have the advantage of immediacy allowing more potential faces to be tried in a short time, and a range of faces consistent with the skull to be quickly produced. These are first presented as electronic images, but may be automatically generated as solid models using numerically controlled milling or rapid prototyping machines.

More automatic methods of identifying corresponding points on facial surface and skull for registration and improved mathematical techniques for modelling the skull itself, and warping the mesh or collection of points representing the facial surface model, are being developed.[19]

The shape of the face of a living person is determined by both bony and cartilaginous structures. Since the bones of the skull alone do not carry enough information to determine the shape of the face, the reconstruction derived from the underlying bone structure is expected to bear only a resemblance to that of the actual face. Facial reconstructions therefore remain inherently inaccurate, but in forensic applications often prove useful in the identification of found skulls, since they provide enough information for the subject to be recognised.

More developed approaches to this problem of reconstructing a facial surface over a skull have been published.[20] The study by Quatrehomme and colleagues[21] included a method for assessing the accuracy with which the facial surface was reconstructed. The heads of two cadavers were CT scanned and three-dimensional numerical models of the skulls and faces were produced. A semi-automatic method was used to select and register craniometric and cephalometric points, and to deform the facial surface so that the distances between these points corresponded to mean standardised values of tissue thickness. The registration algorithm used was based on salient lines of the skull, called crest lines, rather than single anatomical points. The full development of these methods will eventually lead to a reduction in the subjectivity and a greater repeatability of facial reconstructions. The use of a less sparse representation of the skull shape than that of isolated landmarks also has the potential to produce more accurate results.

Attardi and colleagues have utilised computer tomography to create a 3D soft-tissue and skull model, and then approached the problem of reconstructing the facial surface on a mummy skull by first finding the warp function necessary to deform the model to this skull.[22] The same warping function is then applied to the soft tissue of the model. It is argued that if the model skull type used is similar to that of the mummy skull under investigation, then the most accurate reconstruction is likely to result from this procedure. A semi-automated interactive procedure is used to adjust the facial surface to the skull base on algorithms written for computer animation.[23]

Other methods have generated average skull wire-frame templates from CT data, as well as facial (soft-tissue) wire-frame templates from MRI data. The skull templates are warped to match the scans of found skulls or mummy skulls. The warp used determines that which is then applied to the average facial surface template to reconstruct a face consistent with the skull under investigation. As more individuals are added to their MRI database, a greater variety of average soft-tissue and skull templates becomes available for recovered individuals of various combinations of ethnic group, sex, and age.

Creating the 3D Portrait

Fortunately, in the tradition of her group, Hermione's portrait was painted on her coffin. This coffin portrait provided the information required to create the most realistic impression of Hermione's face. The portrait was digitised using a scanner at a resolution of 600 dots per inch. The reconstructed 3D facial surface was converted into a polygonal dataset. The digitally enhanced portrait was then texture-mapped onto the reconstructed

[18] Vanezis et al. (1989).

[19] Quatrehomme et al. (1997); Attardi et al. (1999).

[20] Quatrehomme et al. (1997); Attardi et al. (1999).

[21] Quatrehomme et al. (1997).

[22] Attardi et al. (1999).

[23] Beier and Neely (1992).

Figure 15.7 Coffin portrait, and texture mapping onto reconstructed face.

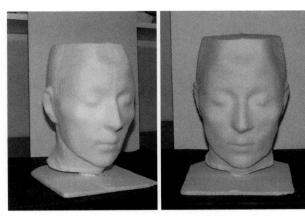

Figure 15.8 Bust of Hermione milled in dense polyurethane.

facial surface, with the main features of the portrait face scaled and brought into registration with the reconstructed 3D facial surface. The software used for this was a leading commercial package called Lightwave (NewTek Inc.) normally used for ray tracing and special effects for the film industry. The polygonal dataset along with its texture map may now be treated as an independent object that may be animated and ray-traced to produce a series of three-dimensional images of Hermione's face. It may also be presented as an animated video on a monitor screen and viewed from any chosen direction. A single view along with the coffin portrait is shown in Figure 15.7.

Creating a Solid Model

The geometry of the derived surface of the head and face of Hermione was used to create a datafile in a suitable format to drive a numerically controlled milling machine. This technique has been developed for clinical purposes to make surgical models of individual anatomies for the manufacture of customised prostheses and implants, and is described elsewhere.[24] A solid model of Hermione's head was milled in dense polyurethane. Photographs of this model are presented in Figure 15.8. The model was shown along with the visualised reconstructions in the 'Ancient Faces' exhibition held at the British Museum. The ability to produce solid models in this manner is of great potential use to museums since the facility makes it possible for a number of museums to have solid replicas of mummy material and other archaeological objects.

Discussion

Although the reconstruction of the bones and remains of soft tissue by the techniques described here will be accurate, the final question remains as to how accurate a representation of the face has been achieved.

Even with the application of more powerful mathematical methods, since the skull does not carry enough information to completely determine the facial surface,

facial reconstruction from skulls will always contain an element of art. The reconstructed face is likely to resemble that of the living person but is very unlikely to be an exact replica.

The basic facial shape is likely to come close to that of the living person, but the shapes of such features as the nose and lips will be less predictable. The ears are a matter of guesswork, unless, as in the case of Hermione, a portrait or other evidence exists.

Research continues into the extent to which the nose can be reconstructed. Macho, for example, has examined in some detail the degree to which the underlying skull shape determines the morphology of the nose.[25] Multivariate analysis on the morphometric dimensions of the external nose, including the thickness of the soft tissues, and craniometric measurements on large samples of male and female subjects showed that nasal height and nasal length are best predicted by the dimensions of the skull, but that nasal depth and the thickness of the soft tissues are greatly influenced by age. Nasal height and nasal length were shown to be strongly influenced by the height of the bony nose and the prominence of the ossa nasalia. For the nose, these studies led Macho to conclude that obtaining soft-tissue thicknesses alone is not sufficient for successful facial reconstruction, but that a more holistic approach should be used to elucidate the relationships between soft-tissue cover and the underlying hyaline and bony structures.

In spite of its known uncertainties, the reconstruction of the faces of Egyptian mummies captures the attention and fires up the imagination of the public, and it may be claimed that this aspect gives it a recognised place in archaeological studies.

Note

The authors would like to express their thanks to Joyce Filer of the British Museum, London, for allowing the reproduction of some of the photographs in this paper, and also to Dr Robin Richards for the use of the

24 Joffe et al. (1999).

25 Macho (1986, 1989).

computer programs he has written for the facial surface reconstructions.

References

ATTARDI, G., BETRÒ, M., FORTE, M., GORI, R., GUIDAZZOLI, A., IMBODEN, S. and MALLEGNI, F. (1999), '3D facial re-construction and visualization of ancient Egyptian mummies using spiral CT data', in M. A. Alberti, G. Gallo, and I. Jelinek (eds.), *Short Papers and Demos, Eurographics '99*: 134–6.

BALDOCK, C., HUGHES, S. W., WHITTAKER, D. K., DAVIS, R., TAYLOR, J., SPENCER, A. J., SOFAT, A. and TONGE, K. (1994), '3-D reconstruction of an ancient Egyptian mummy using X-ray computer tomography', *Journal of the Royal Society of Medicine* 87/12: 806–8.

BEIER, T. and NEELY, S. (1992), 'Feature-based image metamorphosis', *Computer Graphics* 26/2: 35–42.

BRIER, T. (1998), *The Murder of Tutankhamen* (New York).

FALKE, T. H. M., ZWEYPFENNING-SNIJDERS, M. C., ZWEYPFENNING, R. C. V. J. and JAMES, A. E. (1997), 'Computer tomography of an ancient Egyptian cat', *Journal of Computer Assisted Tomography* 11: 745–7.

FILER, J. (1997), 'Revealing Hermione's secrets', *Egyptian Archaeology: the Bulletin of the Egypt Exploration Society* 11: 32–4.

GRAY, P. H. K. and DAWSON W. (1968), 'Mummies and human remains', *Catalogue of Egyptian Antiquities in the British Museum*, vol. 1 (London): 8, plates va, xxivabc.

HARRIS, J. E. and WENT, E. F. (1980), *An X-ray Atlas of the Royal Mummies* (Chicago).

HARWOOD-NASH, D. C. F. (1979), 'Computer tomography of ancient Egyptian mummies', *Journal of Computer Assisted Tomography* 3: 768–73.

ISHERWOOD, I. and HART, C. W. (1992), 'The radiological investigation', in A. R. David and E. Tapp (eds.), *The Mummy's Tale* (London): 100–11.

JOFFE, J. M., NICOLL, S. R., RICHARDS, R., LINNEY, A. D. and HARRIS, M. (1999), 'Validation of computer-assisted manufacture of titanium plates for cranioplasty', *International Journal of Oral and Maxillofacial Surgery* 28: 309–13.

LEWIN, P. K. and HARWOOD-NASH, D. C. (1977), 'X-ray computed axial tomography of an ancient Egyptian brain', *IRCS Medical Science* 5: 78.

LINNEY, A. D. and ALUSI, G. H. (1998), 'Clinical applications of computer aided visualization', *Journal of Visualization* 1/1: 95–109.

LINNEY, A. D., CAMPOS, J. and RICHARDS, R. (1997), 'Non-contact anthropometry using projected laser line distortion: three dimensional graphic visualisation and applications', *Optics and Lasers in Engineering*, 28/2: 137–55.

MACHO, G. A. (1986), 'An appraisal of plastic reconstruction of the external nose', *Journal of Forensic Sciences* 31/4: 1391–1403.

MACHO, G. A. (1989), 'Descriptive morphological features of the nose: an assessment of their importance for plastic reconstruction', *Journal of Forensic Sciences* 34/4: 902–11.

MANHEIN, M. H., LISTI, G. A., BARSLEY, R. E., MUSSELMAN, R., BARROW, E. and UBELAKER, D. H. (2000), 'In vivo facial tissue depth measurements for children and adults', *Journal of Forensic Sciences* 45/1: 48–60.

MARX, M. and D'AURIA, S. H. (1988), 'Three-dimensional CT reconstructions of an ancient human Egyptian mummy', *American Journal of Roentgenology* 150: 147–9.

MIDDLETON, A. P., LANG, J. and DAVIS, R. (1992), 'The application of xeroradiography to the study of museum objects', *Journal of Photographic Science* 40: 34–41.

NEAVE, R. A. H. (1979), 'Reconstruction of the heads of three ancient Egyptian mummies', *Journal of Audiovisual Media in Medicine* 2: 156–64.

PETRIE, W. M. F. (1898), *Deshasheh 1897 with a Chapter by F. Ll. Griffith* (London).

PHILLIPS, V. M. and SMUTS, N. A. (1996), 'Facial reconstruction: utilization of computerized tomography to measure facial tissue thickness in a mixed racial population', *Forensic Science International* 83: 51–9.

PICKERING, R. B., CONCES, D. J., Jr, BRAUNSTEIN, E. M. and YURCO, F. (1990), 'Three-dimensional computed tomography of the mummy Wenuhotep', *American Journal of Physical Anthropology* 83: 49–55.

PRAG, J. and NEAVE, R. (1997), *Making Faces* (London and Texas).

QUATREHOMME, G., COTIN, S., SUBSOL, G., DELINGETTE, H., GARIDEL, Y., GRÉVIN, G., FIDRICH, M., BAILET, P. and OLLIE, A. (1997), 'A fully three-dimensional method for facial reconstruction based on deformable models', *Journal of Forensic Sciences* 42/4: 649–52.

RHINE, J. S. and CAMPBELL, H. R. (1980), 'Thickness of facial tissues in the American Blacks', *Journal of Forensic Sciences* 25: 847–58.

RHINE, J. S., MOOER, C. E. and WESTON, J. T. (1982), *Facial Reproduction: Tables of Facial Tissue Thickness of American Caucasoids in Forensic Anthropology*, Maxwell Museum Technical Series No. 1 (Albuquerque).

SUZUKI, K. (1948), 'On the thickness of the soft parts of the Japanese face', *Journal of the Anthropological Society of Nippon* 60: 7–11.

VAHEY, T. and BROWN, D. (1984), 'Comely Wenuhotep: computer tomography of an Egyptian mummy', *Journal of Computer Assisted Tomography* 8: 992–7.

VANEZIS, P., BLOWES, R. W., LINNEY, A. D., TAN, A. C., RICHARDS, R. and NEAVE, R. (1989), 'Application of 3-D computer graphics for facial reconstruction and comparison with sculpting techniques', *Forensic Science International* 42: 69–84.